Constructing Social Research

Sociology for a New Century

A PINE FORGE PRESS SERIES

Edited by Charles Ragin, Wendy Griswold, and Larry Griffin

Sociology for a New Century brings the best current scholarship to today's students in a series of short texts authored by leaders of a new generation of social scientists. Each book addresses its subject from a comparative, historical, global perspective, and, in doing so, connects social science to the wider concerns of students seeking to make sense of our dramatically changing world.

- *How Societies Change* Daniel Chirot
- *Cultures and Societies in a Changing World* Wendy Griswold
- *Crime and Disrepute* John Hagan
- *Racism and the Modern World* Wilmot James
- *Gods in the Global Village* Lester Kurtz
- *Constructing Social Research* Charles C. Ragin
- *Women, Men, and Work* Barbara Reskin and Irene Padavic
- *Cities in a World Economy* Saskia Sassen

Forthcoming:

- *Social Psychology and Social Institutions* Denise and William Bielby
- *Global Transitions: Emerging Patterns of Inequality* York Bradshaw and Michael Wallace
- *Schools and Societies* Steven Brint
- *The Social Ecology of Natural Resources and Development* Stephen G. Bunker
- *Ethnic Dynamics in the Modern World* Stephen Cornell
- *The Sociology of Childhood* William A. Corsaro
- *Waves of Democracy* John Markoff
- *A Global View of Development* Philip McMichael
- *Health and Society* Bernice Pescosolido
- *Organizations in a World Economy* Walter W. Powell

Constructing Social Research

The Unity and Diversity of Method

Charles C. Ragin
Northwestern University

PINE FORGE PRESS
Thousand Oaks ◆ London ◆ New Delhi

For information, address:

Pine Forge Press
A Sage Publications Company
2455 Teller Road
Thousand Oaks, California 91320
(805) 499-4224
Internet:sdr@pfp.sagepub.com

Production: Scratchgravel Publishing Services
Designer: Lisa S. Mirski
Typesetter: Scratchgravel Publishing Services
Cover: Lisa S. Mirski
Print Buyer: Anna Chin

Printed in the United States of America

99 10 9 8 7 6 5

Library of Congress Cataloging-in-Publication Data

Ragin, Charles C.
Constructing social research : the unity and diversity of method / Charles C. Ragin.
p. cm. — (Sociology for a new century)
Includes bibliographical references and index.
ISBN 0-8039-9021-9 (pbk. : alk. paper)
1. Social sciences—Research. I. Title. II. Series.
H62.R23 1994
300'.72—dc20 93-47611
CIP

Contents

ABOUT THE AUTHOR

Charles C. Ragin spent most of his youth in Texas and the southeastern United States. He attended the University of Texas at Austin as an undergraduate and received his B.A. degree in 1972 at the age of nineteen. That same year he began graduate work in sociology at the University of North Carolina at Chapel Hill, and he received his Ph.D. in 1975. Since 1975, he has lived in the Midwest, teaching first at Indiana University and then at Northwestern University, where he is currently Professor of Sociology and Political Science and also Research Fellow at the Center for Urban Affairs and Policy Research. He is best known for his work on comparative methodology and for his research articles addressing broad issues in politics and society, with topics ranging from the causes of ethnical political mobilization to the shaping of the welfare state in advanced capitalist countries. His book *The Comparative Method: Moving Beyond Qualitative and Quantitative Strategies* won the 1989 Stein Rokkan Prize of the International Social Science Council of UNESCO. He coedits the Pine Forge Series, **Sociology for a New Century,** with Larry Griffin and Wendy Griswold. He is married to Mary Driscoll, and they have two sons, Andrew and Daniel.

ABOUT THE PUBLISHER

Pine Forge Press is a new educational publisher, dedicated to publishing innovative books and software throughout the social sciences. On this and any other of our publications, we welcome your comments, ideas, and suggestions. Please call or write to:

Pine Forge Press
A Sage Publications Company
2455 Teller Road
Thousand Oaks, California 91320
(805) 499-4224
Internet:sdr@pfp.sagepub.com

Foreword

Sociology for a New Century offers the best of current sociological thinking to today's students. The goal of the series is to prepare students, and—in the long run—the informed public, for a world that has changed dramatically in the last three decades and one that continues to astonish.

This goal reflects important changes that have taken place in sociology. The discipline has become broader in orientation, with an ever growing interest in research that is comparative, historical, or transnational in orientation. Sociologists are less focused on "American" society as the pinnacle of human achievement and more sensitive to global processes and trends. They also have become less insulated from surrounding social forces. In the 1970s and 1980s sociologists were so obsessed with constructing a science of society that they saw impenetrability as a sign of success. Today, there is a greater effort to connect sociology to the ongoing concerns and experiences of the informed public.

Each book in this series offers a comparative, historical, transnational, or global perspective in some way, to help broaden students' vision. Students need to be sensitized to diversity in today's world and to the sources of diversity. Knowledge of diversity challenges the limitations of conventional ways of thinking about social life. At the same time, students need to be sensitized to the fact that issues that may seem specifically "American" (for example, the women's movement, an aging population bringing a strained social security and health care system, racial conflict, national chauvinism, and so on) are shared by many other countries. Awareness of commonalities undercuts the tendency to view social issues and questions in narrowly American terms and encourages students to seek out the experiences of others for the lessons they offer. Finally, students also need to be sensitized to phenomena that transcend national boundaries—trends and processes that are supranational (for example, environmental degradation).

Constructing Social Research presents a fresh perspective on the practice of social science. By comparing social research to other ways of constructing representations of social life, it is possible to see the distinctiveness of social research. Unlike other ways of representing social life, such as journalism, social research involves a special dialogue between ideas and evidence. The end result of this dialogue is social scientific representations of social life that are strongly grounded in theory and data. Charles Ragin elaborates this argument with a variety of examples and shows important continuities in three major forms of social research: qualitative research on commonalities, comparative research on diversity, and quantitative research on relationships among variables. These three strategies provide a solid foundation for the study of all social phenomena, from the examination of the complexities of everyday life to the investigation of the power of transnational processes.

Preface

Most books on social research either do too much or too little. Those that do too much present exhaustive inventories of all the things that social researchers do, especially their different ways of collecting data and analyzing evidence. This orientation leaves the impression that social research is a hodgepodge of unconnected activities—that it lacks coherence. Books that do too little tend to present social research in a one-dimensional manner. Typically, social research is presented either as something that is much like research in "hard sciences" like physics, or as something more humanistic in its orientation. Neither portrayal of social research is accurate.

This book does not offer an exhaustive inventory of all the different techniques of data collection and analysis that social researchers use. Books that inventory methods often leave students disoriented. In the end, students can recognize a lot of different "trees," but they are lost in the forest of social research. As an alternative, this text offers a broad, integrative overview of social research, an invitation to social research that interprets the state of the art. It answers the question "What is social research?" with diverse examples that illustrate current thinking about broad issues in social science methodology and the logic of analysis.

The book works best as a starting point for a course in methods of social research or as a capstone to a rigorous or advanced course introducing a social scientific discipline. In courses on methods of social research, it can be used to counterbalance books that present social research in a one-dimensional manner (for example, as an attempt either to approximate or to repudiate research practices in the hard sciences) or to supplement an omnibus text that attempts to present the great variety of methods that social researchers use.

My primary goal in this book is to show unity within the diversity of activities that are called social research—to make sense of social research in a way that brings it all together. In truth, social research *is* diverse. Some social researchers study census data on hundreds of thousands of people; some study one person at a time in an in-depth manner. Some monitor what is happening in the streets today; some try to reconstruct what life was like centuries ago. But there *is* unity in this diversity. As I explain in this book, this unity comes from special features of social research as a way of representing social life. In almost every article, every book, every report that social scientists write, they construct representations of social life. I use this idea as a starting point and then show how all social researchers engage in a dialogue of ideas ("theory") and evidence ("data") when they construct representations of social life. In contrast to many other ways of constructing representations of social life, social research almost always involves a systematic collection and examination of large bodies of evidence.

My emphasis is very different from the typical portrayal of social science as having opposing "quantitative" and "qualitative" wings. The usual idea that follows from this portrayal is that quantitative researchers deal with numbers, like frequencies and percentages, while qualitative researchers deal with experience and meaning—things that can be represented or described best with words, not numbers. But all social researchers must deal with both words and quantities in some way. The difference between quantitative and qualitative research is not numbers versus words, but the simple fact that quantitative researchers typically focus on the links among a smaller number of attributes across many cases when they construct representations of social life, while qualitative researchers focus on the links among a larger number of attributes across relatively few cases in their representations. While very different in orientation, the two kinds of research are similar in that both involve a systematic interplay between ideas and evidence.

In Part II of this book I examine these two strategies in depth as well as a third kind of social research, the comparative approach. In many respects this third approach lies halfway between the two main types. Presenting the comparative approach is important because it helps to break down the apparent barrier between qualitative and quantitative strategies. In other words, it is easier to grasp the unity of social research when a third path, intermediate between the two more common paths, is offered. Like qualitative research, comparative research pays close attention to individual cases; like quantitative research, comparative research focuses directly on differences across cases and attempts to make sense of these differences.

While these three paths are clearly distinct from one another, they are best understood as different ways of conducting a social scientific dialogue of ideas and evidence. In each approach, clear ideas are essential to the process of constructing representations because they help researchers organize and make sense of the large bodies of evidence involved in every social scientific investigation.

I have come to this view of social research gradually, over many years, with lots of help from others. The many who have helped me include the teachers of research methodology I had as an undergraduate (Gideon Sjoberg and Sheldon Ekland-Olson) and as a graduate student (David Heise, Patrick Horan, and Krishnan Namboodiri), and colleagues and friends from my stay at Indiana University (especially Larry Griffin, Kriss Drass, Bill Corsaro, Jacques Delacroix, and David Zaret) and Northwestern University (too many to list, but special thanks to Mary Driscoll, Howard Becker, Arthur Stinchcombe, and Christopher Winship). Steve Rutter and Victoria Nelson of Pine Forge Press and my series coeditors, Larry Griffin and Wendy Griswold, have been generous with their support and encouragement during the writing of this book. I owe its completion to their good humor and quiet confidence that I would finish on time, or at least before a reasonable resemblance of "on time" had passed.

Special thanks also must go to the good people who read this book in draft form and offered great advice about how to turn a very rough draft into something more coherent—Howard Becker, Mary Driscoll, Larry Griffin, Scott Long, S. Philip Morgan, Arthur Stinchcombe, and Peggy Thoits. I am indeed fortunate to have such gifted people offer their suggestions.

If any argument in this work seems especially insightful or interesting, it is almost certainly something that has emerged from my multifaceted collaboration with my wife, Mary Driscoll. As time goes on, I have difficulty separating her thinking from mine. Still, I recognize her insights throughout the book, ideas that would not be there without her help. She has been wonderfully supportive and incisively critical, investing her best energies in comments on my drafts. When the pressure of schedules became too great, she became a coauthor and rescued the book's Afterword. Mary also took on far more than her share of child care and other everyday responsibilities. I feel I cannot fail as long as she is my ally. I hope that I repay my debt to her in our day-to-day lives together and enrich her thinking and writing as much as she enriches mine.

Charles C. Ragin

PART I

Elements of Social Research

Too often we take for granted the things that require the most explanation. After all, it's easier to assume we know what something means than it is to explain or define it.

So it is with the term *social research*. We seem to know what social research is because we are exposed to so much of it. There are frequent reports in the media about unemployment, homelessness, teenage suicide, decaying schools, divorce rates, rising ethnic and racial violence, political anarchy, world poverty, democratization, and other topics. These reports draw on the work and ideas of social researchers.

But can social research be clearly distinguished from other kinds of information gathering and presenting? And if so, how? What are the differences between a report of the insights of a streetwise person and those of a social researcher who spends a lot of time studying social life on the same streets? Like many terms, the scope and meaning of social research can be contested.

Part I of *Constructing Social Research* offers several answers to this basic question—What is social research?—and uses this issue to introduce core features of social science. Some accounts of these core features emphasize the distinctive subject matter of social research, for example, the idea that social researchers study society as a whole and not the psyches of individuals, as psychologists do. Other accounts emphasize its distinctive methods, especially the idea that social researchers use analytic techniques that condense information on many observations. This book offers a more encompassing portrait.

Simply stated, social research is one among many ways of constructing "representations" of social life. A novel, for example, is a representation of social life; so is a documentary film. Both "tell about society" (Becker 1986:121–36). Many different ways of representing social life qualify as social research; many do not. By defining social research as *one* of many ways of constructing representations of social life, it becomes clear that social research has a lot in common with many other kinds of work, for example, the work of writers, journalists, and documentary film producers. While the representations constructed by social researchers may be similar to those produced by others, they also have important distinctive features that should not be overlooked.

Chapter 1 critically evaluates several common answers to the question of how social research differs from other ways of representing social life. These include (1) the proposition that social research has a distinctive subject matter—that it offers a special way of understanding "society;" (2) the argument that social researchers use a special language to tell about society; and (3) the idea that social research is distinguished from other ways of telling about society by its use of the scientific method.

The chapter also compares social research with several other ways of constructing representations of social life. Sometimes it is difficult to tell different kinds of representations apart. For example, when news reporters study issues like homelessness or poverty in a city, their reports focus on much the same factors that social researchers working on the same topic might study. While the similarities between social research and other ways of representing are striking, several important features, as we shall see, distinguish social research. These features reflect the goals of social research and the impact of these goals on the construction of social scientific representations.

Chapter 2 examines the diverse goals of social research, with a special focus on the tensions among different goals. Too often social research is portrayed simply as a process of testing general ideas, or theories, and producing broad statements, or generalizations (Hoover 1976). That is, social research is portrayed as a "hard science" like physics or chemistry, the main difference being its focus on a distinctive and difficult subject matter—social life. However, many social researchers pursue goals that are at odds with the goals of testing theory or producing generalizations. For example, some researchers offer new interpretations of historical events; others seek to "give voice" to marginal groups in society; still others try to pinpoint the cultural significance of new trends. The current diversity of goals should be both acknowledged and examined because different research goals often lead to very different strategies of social research.

Chapter 3 addresses general features of the *process* of social research—how social researchers construct representations of social life. Social research, like almost all research, is organized and systematic, and social researchers usually must follow a plan that helps them make sense of the phenomena they study. For example, social researchers typically collect a lot of evidence when they conduct their studies. However, every person, every situation, every event potentially offers an infinite amount of information. Which bits of information should the researcher pay attention to? How is this to be decided? Imagine being interviewed by a social researcher about *everything* you did yesterday. The interview could take weeks to complete.

The more explicit a researcher's initial ideas (or "analytic frame"), the clearer the guidance they offer about what should be studied and what can safely be ignored in the infinity of information that every case and every situation presents. While this guidance is helpful, it also can be limiting and thus restrict the researcher's view. Sometimes, therefore, initial ideas are intentionally underdeveloped, so that the researcher can be more open to new insights. Chapter 3 argues that the interplay of ideas and evidence is common to all strategies of social research; however, the nature of this interplay can differ substantially from one strategy to the next.

1

What Is Social Research?

Introduction

There are many ways to study and tell about social life. Sometimes it is hard to tell which of these are social research and which are not. Consider a few examples.

Peter Evans spent a lot of time talking to business executives and government officials in Brazil and wrote a book about it called *Dependent Development* (1979). On the basis of his interviews and other work, he found that Brazil's top economic and political leaders were closely enmeshed with some of the world's most powerful multinational corporations, many of which are based in the United States (Union Carbide, for example). He concluded that this "triple alliance" of the Brazilian government, Brazil's economic elites, and multinational corporations shaped Brazil's industrial development in a way that skewed its rewards toward the rich and powerful in Brazil. The bulk of the population benefited only marginally, if at all, and were subjected to intense government repression.

Arlie Hochschild wanted to understand the "commercialization of human feeling." Many jobs in today's economy require what she calls "emotion work"—employees' use of their own feelings to create an outward, public display that supports a particular image (such as friendliness or helpfulness). Such management of a person's emotions can be used to achieve specific ends, especially in service jobs involving interaction with customers or clients. Hochschild studied a lot of different occupations, but devoted special attention to flight attendants and found that emotion work is an essential part of their work. For example, emotion work is often required to keep unruly and sometimes angry passengers in check. Hochschild summarized what she learned from flight attendants and other kinds of service work in a book called *The Managed Heart: Commercialization of Human Feeling* (1983). She found that there was a tendency for certain kinds of emotion work to be assigned to females, thus encouraging the concentration of women in specific occupations.

Douglas Massey was interested in urban poverty in the United States and wanted to find out why conditions deteriorated so rapidly in African-American, inner-city neighborhoods, especially from the 1970s to the present. He studied the largest cities in the United States and found that cities with the most severe housing segregation by race and income level were the ones that experienced the most severe inner-city deterioration. Essentially, he discovered that the greater the degree of housing segregation, the greater the concentration of increases in poverty in specific inner-city neighborhoods. He reported his provocative conclusions in a book titled *American Apartheid* (Massey and Denton 1993).

These three books address important issues. Why do so many people in Third World countries like Brazil still suffer from serious poverty despite the substantial industrialization that has occurred? Why is it that women more than men are required to perform emotion work in their jobs? What factors reinforce this pattern? Why have so many inner-city neighborhoods in the United States suffered such serious deterioration? These questions and the studies that address them are as relevant to the everyday concerns of the informed public as they are to government officials responsible for formulating public policies. The conclusions of any of these three authors could be reported on a television news or magazine show like "Nightline," "60 Minutes," or the "MacNeil/Lehrer NewsHour." The phenomenon of emotion work could even be the basis for a talk show like the "Oprah Winfrey Show."

At first glance, it might appear that these three books were written by journalists or free-lance writers. Yet, all three were written by social researchers trying to make sense of different aspects of social life. What distinguishes these works as social research? More generally, what distinguishes social research from other ways of gathering and presenting evidence about social life? All those who write about society construct **representations** of social life—descriptions that incorporate relevant ideas and evidence about social phenomena. Are the representations constructed by social researchers distinctive in any way from those constructed by nonsocial scientists, and, if so, how?

At the most general level, **social research** includes everything involved in the efforts of social scientists to "tell about society" (Becker 1986). Both parts of this definition of social research—that it involves a *social scientific way* of *telling about society*—are important.

Telling about *society* has special features and some special problems. These problems affect the work of all those who tell about society—from social researchers to novelists to documentary film makers—and sepa-

---

Note: **Boldface** terms in the text are defined in the glossary/index.

rates those who tell about society and social life from those who tell about other things. Social researchers, like others who tell about society, are members of society. They study members of society, and they present the results of their work to members of society. Thus, at a very general level, social researchers overlap with those they study and with the audiences for their work; and those they study—other members of society—also overlap with their audiences.

Among those who consider themselves scientists, this three-way mixing of researcher, subject, and audience exists only in the social and behavioral sciences (anthropology, sociology, political science, and so on) and has an important impact on the nature and conduct of research. For example, it is very difficult to conduct social research without also addressing questions that are fundamentally interpretive or historical in nature—who we are and how we came to be who we are. It is very difficult to neutralize social science in some way and see studying people the same as studying molecules or ants.

The importance of the other part of the definition—that there is a specifically social scientific way of telling—stems from the fact, already noted, that there are lots of people who tell about society. Journalists, for example, do most of the things that social scientists do. They try to collect accurate information (data); they try to organize and analyze the information they gather so that it all makes sense; and they report their conclusions in writing to an audience (typically, the general public). Do journalists conduct social research? Yes, they often do, but they are not considered social scientists. It is important to contrast social research with a variety of other activities so that the special features of the social scientific way of representing social life are clear.

The main concern of this chapter is what is and what isn't social research. I first examine conventional answers to the question of the distinctiveness of social research. Most of these conventional answers are too restrictive—too many social researchers are excluded by these answers. Next, I compare social research to some other ways of telling about society to illustrate important similarities and differences. Too often social researchers are portrayed as ivory tower academics poring over their computer printouts. In fact, social researchers are quite diverse. Some have a lot in common with free-lance writers; others are more like laboratory scientists. Finally, I argue that it is important to focus on how social researchers construct their representations of social life for their audiences, especially for other social scientists. By examining the nature of the representations that social researchers construct, it is possible to see the distinctive features of social research—the social scientific way of representing social life.

Some Conventional Views of Social Research

There are three conventional answers to the question "Does social research constitute a distinctive way of telling about society?" The first argues that social scientists have a special way of defining *society* and this makes social research distinctive. The second asserts that social research relies heavily on the language of *variables* and *relationships among variables* and that this special language sets social scientists apart. The third emphasizes the use of the *scientific method* and the consequent similarities between the social sciences and hard sciences like physics and chemistry.

Do Social Researchers Have a Special Way of Defining Society?

One reason social research has so many close relatives, like journalism and documentary film making, is that many different kinds of work involve telling about society. It's important to unpack the phrase "telling about society" because it might be possible to distinguish social researchers from others who tell about social life and social events by giving the term *society* a special meaning for social researchers or by showing that social scientists all use "society" in a special way.

Society could be used to refer to all inhabitants of a nation-state (for example, all people living in Peru or in the United States). *Telling about society* thus would mean that social research involves making statements about whole countries. For example, a social researcher might show that Americans are more acquisitive or more tolerant than people in other countries. Another might show that the occupational rewards for educational achievement are better in Germany than in most other advanced countries. To understand social research in this way is to see countries as the fundamental unit of social scientific knowledge.

The problem with this way of restricting the definition of social research is that very few of the people who call themselves social researchers make statements that are so broad. Some social researchers study the social relations of a single individual. For example, in *Working Knowledge*, Douglas Harper (1987) examined the social world of a single rural handyman (see also Shaw 1930). Even those who examine whole countries readily admit that in every country there is great social diversity—that many different "social worlds" exist side by side, entwined and overlapping.

Social researchers also acknowledge that they don't have a good working definition of society (see Marsh 1967 for an attempt to offer one). When U.S. citizens visit Canada for an extended period, are they no longer members of "U.S." society? Is there a separate Canadian society or only a single American society, embracing both Canada and the United States? And what about Quebec? What about native Americans? While it is tempting to equate nation-states and societies, and many social scientists routinely do this, it is a hazardous practice. Most of the things that might be called societies transcend national boundaries.

Alternatively, society might be restricted to *formal properties* of human organization and interaction. A **formal property** is a generic feature or pattern that can exist in many different settings. When only two people interact, they form a dyad; when three people interact, they form a triad; and so on. As sociologist Georg Simmel (1950) noted a long time ago, dyads and other basic forms of association have special features regardless of where they are found. This is what makes them "formal" or "generic" properties.

For example, forming a business partnership with another person, a dyad, has a lot of the same qualities as getting married, another dyad. The relationship is both intense and fragile, and typically involves many mutual obligations and rights. Another formal property is size: interaction patterns are different in small and large groups, regardless of setting. Hierarchy—the domination of the many by the few—is another key feature of human social life (Michels 1959). Organizations and groups that are more hierarchical differ systematically from those that are less hierarchical, again, regardless of setting.

While formal properties are important, and almost no one other than social researchers studies them in depth, the investigation of formal properties today constitutes only a relatively small portion of all social research. Many of the things that interest social researchers and their audiences are important not because of their generic features like their size or their degree of hierarchy, but because of their historical or cultural significance.

It is of special importance to Americans, for example, that some hierarchies overlap with racial differences. One overlap is in education: for a variety of reasons, a smaller percentage of African Americans attend college than whites. Such overlapping hierarchies are historically rooted, and they are the focus of frequent and intense political debate. These and many other topics of great importance to social researchers and their audiences cannot be addressed as generic features of human social organization. It is difficult to neutralize their social and political significance, to sanitize them and treat them as abstract, formal properties.

What Is Society?

Society is best understood as *social life*, which, in turn, can be understood in simple and conventional terms as *people doing things together* (Becker 1986). Telling about society basically involves studying how and why people do things together. They make and unmake families and firms; they join and leave neighborhoods and churches; they resist authority; they form political parties and factions within them; they go on strike; they organize revolutions; they make peace; they have fun; they rob gas stations. Historical events and trends (for example, the Civil War in the United States or declining rates of childbearing in nineteenth-century France) are examples of people doing things together. The list is endless. People doing things together is sometimes history making; more often, it is ordinary, everyday, unrecorded social life. Social scientists study all kinds of social activity. Some prefer to study the ordinary; others prefer to study the momentous.

While it may seem contradictory, the category "people doing things together" also includes people *refusing* to do things together (see Scott 1990). For example, when someone decides not to vote in an election because he or she dislikes all the candidates or is disillusioned with the whole electoral process, a nonaction (that is, not voting) has a social character. Not voting, in this light, is intentional and thus can be viewed as an accomplishment. It has a clear and interpretable basis and meaning in everyday social life.

Similarly, the condition of apathy—generally considered to be a passive, empty state—may likewise be a social accomplishment. People are often in situations where they must explain why they don't care about the things they say don't concern them. For example, a person may claim to be unconcerned about environmental pollution in Eastern Europe because the problem is "not close to home." This rationalization can be seen as "apathy work" and the resulting apathy as an accomplishment. As such, it has a clear, definable basis in social life.

Many refusals are clear acts of defiance (Scott 1976, 1990). The prison inmate who starves himself to protest inhuman conditions may seem contradictory or self-destructive, but his body may be his only possible arena for self-assertion in a setting that imposes such severe restrictions. Similarly, the private act of writing in a diary is a refusal to share important personal information, or at least information that is made important by turning it into a secret record. This type of refusal is rooted in the everyday experiences of those who seek this form of privacy.

Even the act of suicide, which at first glance seems very personal and individual, is the ultimate refusal to do things together and thus falls well within the purview of social research. Emile Durkheim (1951), an early French sociologist, was one of the very first social scientists to argue that such refusals are inherently social. They have social causes, social consequences, and social meaning.

The category "people doing things together" and its companion category "refusals" together encompass a broad range of phenomena. This breadth is necessary because a close examination of the work of social researchers shows that their topics are diverse and almost unbounded. This working definition of society does little, however, to distinguish social research from other ways of telling about society.

Do Social Researchers Use a Special Language?

Alternatively, it might be possible to distinguish social research from other ways of telling about society by the language that social researchers use when they tell about society (Lazarsfeld and Rosenberg 1955). Some social researchers argue that when they tell about society they use the language of variables and relationships among variables to describe patterns, and that this language distinguishes social research from other ways of telling about society. (This general approach is discussed in detail in Chapter 6.)

For example, a social researcher might argue that the most racially segregated cities in the United States have the worst public schools (or, conversely, that the least racially segregated cites have the best public schools). This statement expresses a **relationship** between two variables, degree of racial segregation and quality of public schools, and implies a causal connection: where there is racial segregation, there is less attention to and concern for the overall quality of public schools. Both of these variables are attributes of cities in the United States.

More generally, a **variable** is some general feature or aspect (like degree of racial segregation) that differs from one case to the next within a particular set (like cities in the United States). Variables link abstract *concepts* with specific *measures*. In the example, the researcher might believe that the key to having good public schools in racially mixed cities is a high level of *interracial interaction*. The **concept** of interracial interaction, like most concepts, is very general and can be applied in a variety of ways to very different settings (for example, countries, cities, shopping malls, bus stops, high schools, and so on). One way to apply this concept to racially mixed cities is through the variable *racial segregation* (the degree to which different races live in their own, separate neighborhoods).

A **measure** is a specific implementation of a variable with relevant data. The measure "percentage of a city's population living in racially homogeneous neighborhoods" is one possible measure of racial segregation. The higher this percentage, the greater the segregation. There are many other, more sophisticated measures of racial segregation (see Massey and Denton 1993).

To see if it is true that the most racially segregated cities have the worst public schools, it would be necessary to measure both variables, the degree of racial segregation and the quality of the public schools, in each city. The quality of public schools might be measured by average scores on standardized tests, graduation rates, or some other measure. Once the two variables are measured, it would be possible to assess the link between them. Is there a correspondence? Is it true that the cities that are more racially integrated have better public schools? Is it true that the worst public schools are in the most racially segregated cities? In other words, do these two features of cities vary together, or "covary"? Social researchers use the term **covariation** to describe a general pattern of correspondence.

Examining the covariation between two features across a set of **cases** (like racial segregation and quality of public schools across U.S. cities) is the most common way of assessing the relationship between two variables. When we say that two variables are related, we are asserting that there is some pattern of covariation. If we found the expected pattern of covariation across U.S. cities (high levels of racial segregation paired with poor public schools and low levels of racial segregation paired with good public schools), then we could say that these two variables covary and we would use quantitative methods (see Chapter 6) to assess the strength of their correspondence. Social researchers calculate *correlations* in order to assess the *strength* of a pattern of covariation.

Just because two variables covary across a set of cases does not *necessarily* mean that one is the **cause** of the other. However, a pattern of systematic covariation can be offered as evidence in support of the idea or proposition that there is some sort of causal connection between them. The language of variables and relationships among variables provides a powerful shorthand for describing general patterns of correspondence. In this example, evidence on many cities can be condensed into a single number, a **correlation**, describing the strength of the covariation between two measures (see Chapter 6).

It is true that the language of variables and relationships among variables peppers the discourse of most social research. However, there are many who do not use this language. For example, a researcher might chart the history of a declining public school system and include consideration of the impact of racial segregation and other racial factors without resorting directly to the language of variables and relationships. This examination would focus on the unfolding of events—who did what and when, why, and how.

Similarly, systematic observation (that is, fieldwork) in a single, failing school might be the focus of another social researcher's investigation. This work, like the historical study, might not entail explicit use of the language of variables and relationships. Instead, it might center on an effort to uncover and represent "what it's like" to be a student or a teacher at this school. This understanding, in turn, might help a great deal in efforts to link racial segregation and the quality of public schools.

Some social researchers try to avoid using the language of variables and relationships among variables altogether. They believe that this language interferes with their attempts to make sense of social life, especially when the goal of the research is to understand how something came to be the way it is (that is, conduct research on historical origins) or to understand something as an experience (that is, conduct research on how people view their lives and their social worlds).

While some social scientists avoid using the language of variables, many *non*social scientists use it regularly. Social researchers do not have a monopoly on the understanding of social life through variables and their relationships. Many journalists use this language, for example, when they discuss differences from one situation to the next or when they talk about social trends and problems. For example, a journalist discussing a recent outbreak of violence in a major city might note that cities with more serious drug problems also have higher rates of violent crime. Policy makers and others who routinely consume the writing of social scientists also use this language. Even politicians and ministers use it, especially when they warn of dark days ahead or the current trends that are ushering in unwanted or dangerous changes.

It also must be admitted that the language of variables and relationships among variables is not a special language. This way of describing social life crops up often in everyday life. For example, we may say that we learn more in smaller classes, or that we enjoy athletic events more when the game is close, or that families living in rural areas are more closely knit, or that local politicians address real issues while national politicians address made-for-TV issues. In each example, two variables are related. The first, for instance, argues that how much students learn (a variable that can be quantified with objective tests) is influenced by another measurable variable, class size. This way of describing and understanding social life is in no way the special province of social scientists or social research.

Does the Scientific Method Make Social Research Distinctive?

The third conventional answer to the question of what makes social research distinctive is the idea that social researchers follow the "scientific method" while most of the others who tell about society, like journalists, do not. This answer makes social research seem a lot more like research in the hard sciences like physics. Progress in these fields is driven primarily by **experiments**, often conducted in laboratories. If social research can claim to follow the same general scientific plan as these "hard" sciences, then it gains some of their legitimacy as purveyors of scientific truths. At least this is the thinking of those who argue that the use of the scientific method distinguishes social research from other ways of telling about society.

The core of the scientific method concerns the formulation and **testing** of hypotheses. A **hypothesis** is best understood as an educated guess about what the investigator expects to find in a particular set of evidence. It is an "educated" guess in the sense that is based on the investigator's knowledge of the phenomena he or she is studying and on his or her understanding of relevant ideas or *social theories* (see discussion of social theory later). Social researchers often develop hypotheses by studying the writings and research of other social scientists. These writings include not only research on a given topic, but also relevant theoretical works. Social scientists use these writings in combination with whatever they know or can learn about their research subject to formulate hypotheses. Hypotheses are most often formulated as propositions about the expected relationship between two or more variables across a particular set or **category** of cases.

Generally, a hypothesis involves the **deduction** of a specific proposition or expectation from a general theoretical argument or perspective. It is a mental act, based on existing knowledge. For example, a researcher might be interested in the impact of occupation on voting behavior, especially the political differences between industrial workers who interact only with machines compared to those who must interact with other workers to coordinate production. In addition to the many studies of voting behavior, the researcher might also consult Karl Marx's (1976) ideas about work and class consciousness presented in his three-volume work *Das Kapital*, Max Weber's (1978) ideas about social class in *Economy and Society*, and the ideas of modern scholars such as Seymour Lipset (1982) and Erik Wright (1985). After consulting all the relevant studies and theoretical writings, the researcher might derive a specific hypothesis: Industrial workers who interact more with machines vote less often than industrial workers who interact with other workers on the job, but when they do vote, they vote more consistently for the Democratic Party.

After formulating a hypothesis, social researchers collect data relevant to the hypothesis and then test the hypothesis with the data they have collected. The test usually involves an examination of patterns in the data to see if they match up well with the patterns predicted by the hypothesis. Analysis of the data may refute or support the hypothesis. Typically, analysis of the data also suggests revisions of the hypothesis that could be explored in a future study.

Information to test the hypothesis could be collected in a variety of ways (for example, telephone interviews, mailed questionnaires, and so on). Once collected, the researcher could use statistical methods to test the hypothesis. The researcher would compare the two categories of industrial workers with respect to their different voting histories—how often they voted and who they voted for—to see if there are substantial differences between the two groups in the ways predicted by the hypothesis.

The examination of the data has important implications for the ideas used to generate the hypothesis. On the basis of the newly collected evidence, for example, the researcher might conclude that these ideas need serious emendation. The use of evidence to formulate or reformulate general ideas is called **induction**. Induction is a process whereby the implications of evidence, especially new evidence combined with existing evidence, for general ideas are assessed.

In the **scientific method**, deduction and induction work together. The hypothesis is derived from theory and from existing knowledge about the research subject. Data relevant to the hypothesis are assembled or collected, and the correctness of the hypothesis is assessed. The new knowledge that is generated through these efforts can then be used, through the process of induction, to extend, refine, or reformulate existing ideas. In short, deduction starts with general ideas and applies them to evidence; induction starts with evidence and assesses its implication for general ideas.

The scientific method dictates that researchers follow specific steps:

- study the relevant literature
- formulate a hypothesis
- develop a research design
- collect data
- analyze the data in a way dictated by the hypothesis

The data support or refute the hypothesis. The scientific method works best when different theories can be used to deduce competing hypotheses. When diametrically opposed hypotheses are deduced from two or more theories, then the analysis of relevant data provides a decisive or "critical" test of opposing arguments. Both theories can't be supported by the same data if they make opposite predictions.

For example, if one theory predicts that national economies subject to *more* government regulation (rules and restrictions on what businesses can do) should have higher economic growth rates when world trade slumps, and a second theory predicts that national economies subject to *less* government regulation should fare better under these conditions, then examination of relevant data on national economies should permit a decisive test of these competing arguments.

While there are many social researchers who use the scientific method as described here, there are also many who do not. For example, some social scientists (see, for example, Smith 1987) believe that the most important thing a social scientist can do is to *give voice* to **marginalized groups** in society—to tell the stories of those who have been shoved aside by the rest of society (see Chapter 2).

Douglas Harper (1982) spent months riding freight cars with "tramps," talking with them and taking photos, in order to build a representation of their lives. The greater the role of preexisting theories and ideas in a project of this sort, the more the voices of the research subjects are blocked by the trappings of hard science imposed on an elusive social phenomenon. The voices of the subjects are lost as the loudspeaker of social science theory drowns out all competitors. This reasoning is inconsistent with the logic of the scientific method which emphasizes the testing of hypotheses.

It is also worth noting that it is not easy to follow the scientific method in social research, even when it is the goal of the researcher to adhere strictly to this framework. Most social scientific theories are abstract, vague, and inconsistent, and it is difficult to deduce clear hypotheses from them. Sometimes a theory is so vaguely formulated that it is possible to deduce contradictory arguments from the same theory.

Furthermore, when analyses of the data used to test a hypothesis do not support it, most researchers are reluctant to conclude that the theory they are testing is wrong. Instead, they usually point to inadequacies in the data, to the impossibility of measuring social phenomena with precision, or to some other practical problem. Finally, social researchers are often known to search their data for interesting patterns, regardless of what was hypothesized. This process of discovery generally makes better use of a data set than strict adherence to the requirements of the scientific method (Diesing 1971).

Like others who tell about society, most social researchers devote their energies to trying to make sense of social life using whatever procedures and strategies seem most useful and appropriate for the questions they address. They worry less about following the strict dictates of the scientific method in their efforts to construct well-grounded representations of social life.

To summarize the discussion of conventional views of the distinctiveness of social research: Social researchers don't have one special way of defining society, at least not that they all agree on. Nor do they have a special way of telling about it that they all agree on. And while many social researchers respect the scientific method, not all follow its dictates strictly, and some ignore its dictates altogether. It *is* true that social researchers have tried harder than others to define society and social life; they *do* tend to use the language of variables and relationships among variables more than anyone else; and many of them *do* test hypotheses according to systematic rules. But these are not *defining features* of social research; they are better seen as *tendencies* of social research.

Social Research and Other Ways of Representing Social Life

Novelists and other writers, journalists, documentary photographers and film makers, and a host of others, in addition to social researchers, construct representations that tell about society. They all address the subtleties of social life—people doing and refusing to do things together. Is it possible to distinguish social researchers from these other ways of telling about society?

Consider documentary film makers first. In some ways, the makers of these films seem more concerned than social researchers with constructing valid representations of social life. When social researchers represent society, they often use tables and charts that condense and simplify the vast amount of evidence they have collected. When a researcher states, for example, that people with more education tend to be more politically tolerant, the conclusion may summarize information on thousands of people canvassed in a survey. Or social researchers may select a quote or two to illustrate a conclusion based on an analysis of hundreds of hours of taped, face-to-face interviews. In almost all social scientific representations of social life, the social researcher explains in detail his or her *interpretation* of the evidence used in the representation.

Documentary film makers, by contrast, try to present much of their evidence up front, often without commenting directly on its meaning or significance. While it is true that film makers select which clips to show and then arrange them in sequence, the representation itself is made up of actual recordings. Also, many documentary film makers avoid injecting verbal or written interpretations of the evidence that is presented. Thus, while documentary films, like all representations of social life, are constructed in ways that reflect the goals and intentions of their makers, these representations often have a lower ratio of interpretation to evidence, and in most instances they display a higher proportion of all the primary evidence collected than representations produced by social researchers. Viewers of documentary films are sometimes left to draw their own conclusions from the representation. Social researchers, by contrast, usually state their own conclusions openly, and they carefully organize their representations around these clearly stated conclusions.

At the other extreme, consider the work of novelists. Some novelists strive to write stories that are as realistic as possible. They create fiction, but their fictions are believable representations of social life, representations that often strike at the core of what it means to live in a complex social world. Imagine a novelist concerned about race in the South. She bases her novel on her experience of race relations as a child growing up in the deep South in the 1950s. She wants to capture, as much as possible, the essence of what it was like. Much of the book might be based on actual experiences—true events—but much of it might be pure fiction as well, events fabricated by the author. Yet these fictions might do a much better job of capturing the essence of what it was actually like to live in the South during this period than a careful recounting of true events. In short, by creating fictions, the novelist might do a better job of capturing the reality, the true character, of race during this period than she might if she were to present a straight history of relevant childhood events.

At one extreme is documentary film, representations based on recorded slices of social life. At the other extreme is the novel, the creation of insightful fiction. Both ways of representing social life have important strengths that are only rarely found in social research. In some ways, social research may seem impotent when compared to these other, more dramatic ways of representing.

But we really don't expect to find these qualities in social research. We don't expect social researchers to present mounds of data. In fact, the social researcher who simply presents mounds of data is considered a failure because the work is not complete. Nor do we want social researchers to create deliberate fictions to enhance the points they want to make. The social researcher who knowingly presents fictions as truth is considered dishonest and, if discovered, will be accused of violating professional ethics.

From the perspective of most social researchers, the representation of social life offered in a novel is overprocessed compared to social science because the representation goes far beyond the evidence. The representations constructed by social researchers are more processed and condensed than those offered in documentary films and less processed than those created in novels. At least this is the happy median that most social researchers strive for—to go beyond raw data and provide a clear interpretation of the evidence, while stopping well short of fiction.

In this respect, social research is a lot like journalism. Journalists process and condense information about social life, but they also try to avoid manufacturing fictions. Among the many ways of telling about society that could be compared to social research, journalism offers the closest and most fruitful comparison.

Journalism and Social Research: The Similarities

Journalists write about what's going on in society; they represent social life. Most often they report on current events, but they also write stories that offer historical perspectives and in-depth interpretations. Journalists also address major trends and social problems, not just the news of the day, and sometimes these reports are very similar to the research reports of social scientists. Also like social researchers, journalists develop special topic areas: some focus on political events; some on economic trends; some on women's issues; some report on everyday life; some analyze major international events and issues; and so on. Virtually all aspects of social life fall within the purview of journalism. If people will read about a topic, journalists will report on it.

Regardless of topic, journalists all face the same problem regarding "evidence" or "facts." This problem parallels that of social researchers facing "data." Like social researchers, journalists collect an enormous amount of information that could become evidence for a report. They have to decide what is relevant as evidence and then identify the most pertinent bits. This process of gathering and selecting evidence goes hand in hand with developing the focus of the investigation and the report. As the report becomes more of a finished product—as it coalesces in the mind of the journalist as a story—the collection of evidence becomes more focused and more selective. Initial ideas become leads; some leads bear fruit and are pursued vigorously; the story takes shape. Lots of potential evidence and potential stories are left behind.

The same holds true for social research. Social scientists must select from the vast amount of information that social life offers and construct their representations from carefully selected bits and slices. Data collection (that is, the process of gathering evidence) is necessarily selective, and becomes much more so as an investigation progresses. The researcher may start with a few ideas (for example, sensitizing concepts; see Chapter 4) and maybe a working hypothesis or two. These ideas determine the initial data collection efforts. As more is learned about the subject, either through data collection or data analysis, the research becomes more focused and fewer avenues are kept open. As the results take shape in the mind of the investigator, much of what was initially thought to be important may be cast aside as irrelevant.

Both social researchers and journalists find that, in the end, much of the evidence they collected at the start of the investigation was based on false leads, and that they could have been much more efficient in their collection of evidence if only they had known at the start what they learned toward the end of the investigation. The collection of evidence is necessarily selective because potentially there is an infinite quantity of evidence. However, both journalists and social researchers find that in the end they cannot use all the evidence they have collected.

There is great danger in both journalism and social research that follows from this need for *selective* gathering of evidence. Sometimes what may be a false lead is not recognized as such, and it may become the focus or at least an important part of the investigation. False leads pose serious problems in both journalism and social research because they may be biased by accepted knowledge, stereotypes, and common, everyday understandings of social life. For example, there are two common images of the African-American male—the dangerous, inner-city ghetto teenager and the upwardly mobile young professional. As Mitchell Duneier (1992) points out in *Slim's Table*, both of these images are media creations and have little to do with the lives of most African-American men. Research or journalism that uses these images as starting points will fail to arrive at valid representations of the experiences of most African-American males.

Another problem is the simple fact that people questioned or studied by a journalist or a social researcher may unconsciously or deliberately seek to deceive those who study them. Both social researchers and journalists strive to get valid evidence. For journalists, this effort is often described as reporting "just the facts" or at least trying to balance different views of the same facts. Journalists check different sources against each other and maintain constant vigilance in their efforts to detect deception. After all, interested parties may have a lot to gain if their version of "the facts" is accepted by a journalist and then reported as the one true version.

While social researchers are less often the target of outright deception, like journalists they must deal with bias, distortion, and cover-up. For example, while it might seem a simple matter to determine the percentage of homosexuals among adult males in the United States, social researchers have come up with a range of answers, from less than 2% to about 10%. (The more recent studies tend to offer the lower estimates.) There are various reasons for this wide range; one of them is surely people's reluctance to discuss their sexual behavior openly.

"Social facts" can be as elusive as bias-free journalism. Thus, the two fields have comparable obsessions with "truth," or **validity** as it is known to social researchers. For journalism this is expressed by a concern for reporting only verifiable information. Thus, journalists are very concerned with fact-checking and with the authority of their sources of information.

Social researchers' concern for validity is seen in their efforts to verify that their data collection and measurement procedures work the way they claim. Researchers attempting to determine the percentage of homosexuals among adult males in the United States, for example, would have to contend with a variety of threats to the validity of their measurement procedures. People with more varied sex lives, for example, are more likely to agree to talk about their sex lives or to fill out questionnaires on their sexual behavior. This **bias** would surely increase the size of the estimate of the percentage of homosexuals based on survey data. Thus, researchers would have to find some way to address this threat to the validity of their measurement procedures and their estimate of the percentage of homosexuals.

Another similarity between journalists and social researchers is that they must analyze and arrange evidence before they can offer their representations of social life for wider consumption (for example, as news or research reports). As evidence is gathered and selected, the investigator tries to make sense of it. Ongoing analysis of the evidence simplifies the task of what to collect next. Once the gathering and selecting of evidence is complete, the *analysis* of evidence intensifies. A thorough analysis of evidence, in both journalism and social research, is an important preliminary to arranging it for presentation in a report.

When social life is represented, both social researchers and journalists make connections in their data. When a journalist reconstructs the story of a political scandal, for example, connections and timing are crucially important to the representation of the scandal. It matters who said or did what and when. The goal of analysis is to make these connections. In social research, connections are often *causal* in nature. An analysis of a decaying section of a city, for example, might focus on the long-term economic and social forces responsible for the decline.

Journalists analyze their evidence to make sure that the proper connections are made; then they arrange the evidence for presentation in a report. Readers want to know the big picture—the journalist's final synthesis of the evidence—not all the bits of evidence that the journalist collected along the way before arriving at a synthesis. It's the same with social research. It's not possible to include all the evidence the social researcher collected when reporting conclusions. The evidence that is represented in a research report is a select subset of the evidence collected, which of course is a select subset of the vast volume of potential evidence.

The similarities between the work of journalists and the work of social researchers are striking. Of necessity, they both selectively gather evidence relevant to specific questions, analyze it, and then select a subset of the evidence they have gathered for reporting. The report itself is an attempt to construct for the reader the investigator's conclusions regarding the evidence. Evidence is arranged and condensed in a way that illustrates the investigator's conclusions. In effect, the reader is presented with the investigator's arrangement of a fraction of the evidence collected, a small fraction of the potential evidence. Thus, in both social research and journalism representations of social life (the end products of efforts to tell about society) are condensed descriptions structured according to the investigator's ideas. These representations emerge from a systematic dialogue between the investigator's ideas and evidence.

How Social Research Differs

Journalists write for wide audiences, usually for the literate public as a whole. They hope to reach as many people as possible. The primary audience for social researchers, by contrast, is social scientists and other professionals. Many social researchers hope to reach, eventually, the literate public with their findings and their ideas, and some social researchers write for these audiences. But most social researchers expect to reach these general audiences indirectly—through the work of others such as journalists and free-lance writers who use the work and the ideas of social researchers.

The importance of this difference can be seen clearly in the work of social scientists who write for several different target audiences. When their primary audience is social scientists and other professionals they emphasize, among other things, technical aspects of their research and its place in a specific research literature—that is, its relation to the work of others who have researched the same or similar topics. When these same researchers write for the general public, however, they usually skip over technical aspects of the research and the discussion of the work of others (research literatures) and focus instead on the relevance of their own research findings to the concerns of the general public.

The point is not that the nature of the target audience shapes the nature of the representation, although this is certainly an important consideration. Rather, it is pinpointing the distinctiveness of the social scientific way of representing social life. The distinctiveness of the social scientific way of telling about society is most apparent when representations of social life produced *by* social scientists *for* social scientists are examined, especially given the fact that social scientists consider it their professional responsibility to monitor and evaluate the quality of each other's representations. It is important, therefore, to address how social researchers construct these representations.

What makes a representation of social life especially relevant to a social scientist? Briefly, social scientific audiences expect social scientific representations:

- to address phenomena that are socially significant in some way,
- to be relevant to social theory, either directly or indirectly,
- to be based on or incorporate large amounts of appropriate evidence, purposefully collected, and
- to result from some form of systematic analysis of this evidence.

While *some* of these features are found in many journalistic representations of social life, *all* four features are commonly found together in most social scientific representations. Because social scientific representations of social life have these four features, they tend to be better grounded in *ideas* and *evidence* than other kinds of representations. Ultimately, it is their strong grounding in ideas and evidence that makes these representations especially relevant to social scientists.

Social Researchers Address Phenomena That Are Socially Significant

Many of the things that social researchers address are socially significant simply because they are general. Social scientists address all kinds of rates and percentages, for example, used to characterize large numbers of people (the homicide rate, the percentage of voters, and so on), and they study variation in these rates (for example, why some groups murder more than others, why some groups vote more than others, and so on). Sometimes rates and percentages are compared across whole countries (for example, rates of infant mortality in Asian versus Latin American countries). While a single murder might be relevant to theory in some way, common acts are more often studied across large populations, as rates and percentages.

It is not simply generality and the possibility of studying rates that makes phenomena socially significant, however. Some phenomena are significant not because they are common, but because they are rare, unusual, or extreme in some way. A researcher might study a business, for example, that attempts to maintain a completely egalitarian structure, with no one giving orders to anyone else. How do they get things done? Or a researcher might study a country with great ethnic and cultural diversity but little ethnic conflict. How is ethnic competition contained? Another researcher might study a poor immigrant group that assimilated quickly and overcame extreme prejudice while achieving breathtaking economic gains. How did they do it when so many other groups have struggled and failed? Finally, another researcher might study women who dress and pass as men. What do they gain? What do they lose?

These phenomena are worth studying because they are uncommon. However, they are studied not simply because of their interest value, but because they are relevant to how social researchers think about what is more common and thus challenge their basic assumptions about social life.

Social phenomena may also be selected for study because of their historical significance. An understanding of slavery, for example, is vitally important to the understanding and interpretation of race in the United States today. Similarly, an understanding of the relations between the United States and its Latin American neighbors, particularly Mexico and Puerto Rico, is central to an understanding of Hispanic Americans. One key to understanding post–World War II U.S. society is the "Bomb" and other nuclear weapons and the collective perception of their destructive potential. Our thinking about the military and military life in general was strongly influenced by the experience of the Vietnam War and, more recently, the Gulf War. In short, many different aspects of our history have an impact on who we are today. It is difficult to know and understand American society without exploring the impact of its history.

Social Researchers Connect Their Work to Social Theory

Social scientific representations of social life almost always address social theory in some way. A study of homicide rates is relevant to theories of social conflict. A study of women who dress and pass as men is relevant to theories that address gender differences and power. But what is social theory?

Most social scientists participate, in one way or another, in a set of loosely connected, ongoing conversations about abstract ideas with other social scientists and social thinkers. These conversations address basic features and processes of social life and seek to answer enduring questions. These conversations started before any of today's social scientists were born and more than likely will continue long after they have all died. While they often focus on abstract social concepts that have been around a long time (like the concept of equality, or the concept of society), they also shift over time, sometimes taking up new topics (gender and power, for example), sometimes returning to old topics (for example, the degree to which a group's culture can change in the absence of significant changes in material conditions such as level of technology).

These long-term, ongoing conversations provide a background for the development of specific social theories that are spelled out in the research process. A **social theory** is an attempt to specify as clearly as possible a set of ideas that pertain to a particular phenomenon or set of phenomena. Clarity is important because social theory guides research. Sometimes the ideas that make up a theory are expressed clearly at the start of a research project in the form of specific assumptions, concepts, and relationships. Research that seeks to follow the plan of the scientific method needs such clarity from the start. The researcher uses theory as a basis for formulating a specific hypothesis that is then tested with data especially collected for the test.

Sometimes, however, ideas are clarified in the course of the research. This approach is common in research that seeks to use evidence to formulate new ideas. Consider the social researcher who studies something a journalist might study, a new religious cult. More than likely, the researcher will compare this cult to a variety of other cults and in this way show the relevance of the cult to theories of religion. By contrast, a journalist might simply focus on the bizarre or unusual practices that set this cult apart from the rest of society.

The social researcher might also question the label "religious cult." Suppose the cult was also very successful at marketing a particular product, something produced by members of the cult (see Zablocki 1980). Is it a cult or is it a new type of business enterprise? Which set of social theories, those addressing religious cults or those addressing economic organizations, is more useful when trying to understand this group? What are the implications of this group for either set of theories? In most social research, there is a clear *dialogue* with social theory that is an essential part of the research process (see Chapter 3).

Social Researchers Use Large Amounts of Purposefully Collected Evidence

Most social researchers summarize mountains of evidence in the representations they construct. Social researchers tend to incorporate a lot of in-depth information about a limited number of cases (as in much **qualitative research**) or a limited amount of information about a large number of cases (as in most **quantitative research**) in their representations. Either way, they collect a lot of data. When social researchers construct representations, they try to incorporate as much of this evidence as possible, either by condensing and summarizing it or by highlighting the essential features of the cases they study.

The audiences for social research expect representations to summarize large amounts of evidence. In journalism, investigation is often focused on fact checking—making sure that each piece of a story is correct. Social researchers, by contrast, usually focus on the "weight" of the evidence. For example, in survey research, the investigator expects some respondents to make mistakes when they try to recall how they voted in the last election. Such mistakes are not fatal because the investigator is interested primarily in broad tendencies in the data—in the average voter or in the tendencies of broad categories of voters. Do richer respondents tend to vote more often for Republican candidates? Social researchers *do* strive for precision—they try to get the facts right, but when they construct representations, their primary concern is to present a synthesis of the facts that both makes sense and is true to the evidence.

While large amounts of evidence are incorporated into most social scientific representations, it is important to recognize that the evidence that is used is *purposefully collected.* In much social research, investigators put together a specific research design. An investigator's **research design** is a plan for collecting and analyzing evidence that will make it possible for the investigator to answer whatever questions he or she has posed. The design of an investigation touches almost all aspects of the research. The important ones to consider here are those that pertain to social scientists' use of large amounts of purposefully collected evidence. These include:

1. **Data collection technique.** Social researchers use a variety of different techniques: observation, interviewing, participating in activities, use of telephone and other types of surveys, collection of official statistics or historical archives, use of census materials and other evidence collected by governments, records of historical events, and so on. The choice of data collection technique is in large part shaped by the nature of the research question. All these techniques can yield enormous amounts of evidence.

2. **Sampling.** In most research situations, investigators confront a staggering surplus of data, and they often need to devise strategies for sampling the available data. The survey researcher who wants to study racial differences in voting does not need to know every voter's preference, just enough to make an accurate assessment of tendencies. A **random sample** of 1,000 voters might be sufficient. A researcher who wants to study how protest demonstrations have changed over the last twenty years based on an in-depth investigation of fifty such demonstrations must develop a strategy for selecting which fifty to study.

3. **Sample selection bias.** Whenever researchers use only a subset of the potential evidence, as when they sample, they have to worry about the **representativeness** of the subset they use. A study of poor people that uses telephone interviews is not likely to result in a representative sample because many, many poor people (in addition to thousands of homeless people) cannot afford phones. Likewise, the researcher who selects fifty protest demonstrations to see how these demonstrations have changed over the last twenty years must make sure that each one selected is sufficiently representative of the period from which it was selected.

4. **Data collection design.** Sometimes researchers collect a lot of evidence but then realize that they don't have the right kinds of evidence for the questions that concern them most. For example, a researcher interested in the differences between upper income whites and upper income blacks may discover all too late that a random sample of a large population typically will not yield enough cases in these two categories, especially upper income blacks, to permit a thorough comparison. Most issues in data collection design concern the *appropriateness* of the data collected for the questions asked. A study of the impact of a new job training program that provides workers with new skills, for example, should follow these workers for several years, not several weeks or months. The *timing* of data collection (or "observation") is an important issue in almost all studies. More generally, social researchers, more than most others who represent social life, recognize that the nature of their evidence constrains the questions that they can ask of it (see especially Lieberson 1985).

Systematic collection of evidence is important even in research that is more open-ended and less structured from the start of the investigation (as in most qualitative research; see Chapter 4). Often in research of this type, issues of sampling and selection bias are addressed in the course of the research, as the investigator's representation takes shape. A researcher who discovers some new aspect of a group in the course of informal observation will develop a data collection strategy that allows assessment of the generality of the phenomenon (Glaser and Strauss 1967; Strauss 1987).

Social Researchers Analyze Evidence Systematically

The power of the analytic tools social researchers apply to their evidence is sometimes staggering. Powerful computers, for example, are needed to examine the relationship between household income and number of children across the hundreds of thousands of households included in census data banks. Do families with larger incomes have more or fewer children? It's very difficult to answer this question without a computer and sophisticated statistical software. Most social scientific representations result from the application of some systematic technique of data analysis to a large body of evidence. Different procedures for analyzing evidence are used for different kinds of evidence.

Consider the researcher interested in why some women try to dress and pass as men. First, it is clear that to answer this question it would be necessary to interview a substantial number of women who do this. Some effort should be made to talk to women from as many different walks of life as possible. Perhaps women from different ethnic or class backgrounds do it for different reasons. Maybe some are lesbian and some straight, and their reasons differ. It might be necessary to interview thirty to sixty women. Because it is a sensitive topic, and rapport between these women and the researcher is important, these interviews would need to be in depth, perhaps stretching two to four hours each. Assume fifty women are interviewed for 3 hours each. The researcher then would have a total of 150 hours of taped interviews. How can this large body of evidence be shaped into a representation of the social significance and meaning of cross-dressing for these women?

Social scientists have devised a variety of techniques for systematically analyzing this kind of evidence. Most focus on clarifying the concepts and categories that help make sense of this mass of evidence (see Chapter 4). The issue here is not the specific techniques, but the fact that most audiences for social research expect the representation of this kind of evidence to be based on systematic analysis of the entire body of evidence. A journalistic representation, by contrast, might simply tell the stories of a handful of the most interesting cases.

More generally, techniques for the systematic analysis of data are a central part of research design. As noted, the term *research design* embraces all aspects of the collection and analysis of data. Just as most researchers develop a systematic plan for the collection of data—to make sure that they have evidence that is relevant to the questions they ask—they also develop a plan for analyzing their data. In the cross-dressing study, the plan would involve how to make best use of the hundreds of hours of taped interviews. How does one go about identifying commonalities in the things these women said and how they said them? In a very different type of study, say a survey addressing the relationship between social class and attitudes about abortion, the analysis plan would focus on the measurement of the main variables (social class and attitudes about abortion) and different ways of relating them statistically (see Chapter 6).

Conclusion

Social researchers, like many others, construct representations of social life. A study showing that single men are less satisfied with their lives than married men, single women, or married women is a representation of one aspect of society—the complex relations among gender, marital status, and personal satisfaction.

Social researchers construct representations of society and then publish them, usually in scientific journals (for example, *American Sociological Review*, *American Political Science Review*, *American Anthropologist*, and *Journal of Social History*); in scholarly books, reports, and monographs; in textbooks and other teaching material; and sometimes in magazines, newspapers, and trade books—when they want to reach nonacademic audiences. While social scientific representations usually appear in print, they are not limited to these media. They may also be oral (for example, public lectures). They may include tape recordings, photographs, videotapes, documentary films, and even dramatic productions. Thus, social research has a lot in common with other ways of representing social life, but it is also a distinctive way of representing. It is a lot like journalism, but most social research differs in important ways from journalism.

Social research is not for everyone. Many would rather not participate in age-old conversations about fundamental social questions. It's often easier to ignore what other researchers and social thinkers have said. Many consider it tedious to collect large quantities of evidence. It all seems repetitious and painstaking. Many don't want to bother learning how to conduct systematic analysis of large bodies of evidence. After all, it's much easier to find a few easy cases that are interesting and focus on them. Who wants to learn statistics or how to code evidence from hundreds of hours of taped interviews?

It's also true that the evidence itself may seem too constraining. Both journalists and social researchers have trouble with pesky evidence—data that don't give the exact message the investigator would like to present. The social "truths" that can be manufactured through novels, plays, and other forms of fiction may be much more appealing. Finally, some people want their cases to "speak for themselves" as much as possible. They may prefer to present exact recordings like videotapes and let their audiences choose their own messages in these representations.

While social research is difficult and limiting, it also offers special rewards for those willing to make the investments. People who like to read and write about social issues are drawn to social research. Often they have strong political commitments (for example, to fairness in the economic and political arenas). They hope to translate their concerns into publications—representations of social life—that influence social policy. Publications can influence policy directly by bringing issues to the attention of public officials or indirectly by altering the social consciousness of the informed public. Like the three researchers mentioned in the introduction to this chapter, thousands of other social researchers have constructed representations of social life reflecting their concerns. Many have had a direct or indirect impact on social issues.

The beauty of social research is that it tempers and clarifies the concerns and interests of those who practice the craft. Social research has this impact on people who address social issues in several ways: Social researchers must engage in long-standing debates about society and social life when they conduct research. Social researchers must base their representations on systematic examination of large quantities of systematically collected evidence. Social researchers as a community pass judgment on the representations of social life produced by social researchers (Merton 1973; Kuhn 1962). In effect, they inspect and evaluate each other's work.

Thus, of all ways of representing social life, those that emanate from social research have a very strong grounding in ideas and evidence and a great potential for influencing social policy. As a community of scholars, social researchers work together to construct representations of social life that fulfill the many and varied goals of social research, from documenting broad patterns and testing social theories to giving voice to marginal groups in society.

2

The Goals of Social Research

Introduction

Social life is infinitely complex. Every situation, every story is unique. Yet, people make their way through this world of complexity. Most things, most situations seem familiar enough, and people can usually figure out how to avoid the unfamiliar. Also, there is order in complexity, even if people are not always conscious of the order. Some of this order-in-complexity is easy to describe (as in what sports fans do to mark certain events in a game). Other examples of order-in-complexity are difficult to explain, much less describe (for example, the interplay of pagan and Christian symbols in the historical development of an elaborate religious ritual).

Social researchers seek to identify order and regularity in the complexity of social life; they try to make sense of it. This is their most fundamental goal. When they tell about society—how people do or refuse to do things together—they describe whatever order they have found. There is even a describable order to what may appear to be social chaos, such as a mass political demonstration that gets out of hand and leads to a violent attack on nearby symbols of authority.

While identifying order in the complexity of social life is the most fundamental goal of social research, there are many other, more specific goals that contribute to this larger goal. They are quite diverse. For example, the goal of testing theories about social life contributes to the larger goal of identifying order in complexity; so does the goal of collecting in-depth information on the diverse social groups that make up society. Another factor that contributes to the diversity of the goals of social research is the simple fact that social research reflects society, and society itself is diverse, multifaceted, and composed of many antagonistic groups. It follows that the goals of social research are multiple and sometimes contradictory. Today, no single goal dominates social research.

Several of the main goals of social research resemble the goals of research in the "hard" sciences like physics and chemistry. These goals include, for example, the identification of general patterns and relationships. When we show that people with more education tend to vote more often and that this link exists in many democratic countries, we have documented a general relationship for individuals living in democracies. Similarly, when we observe that countries with greater income inequality tend to be more politically unstable, we have identified a pattern that holds across entire nation-states.

Knowledge of general patterns and relationships is valuable because it is a good starting point for understanding many specific situations and for making predictions about the future. Also, general patterns in society are directly relevant to the testing of social science theory—the body of ideas that social scientists often draw upon in their efforts to make sense of and tell about society.

Some of the other goals of social research, however, are not modeled on the hard sciences. These other goals follow more directly from the fact that social researchers are members of the social worlds they study (see Chapter 1). For example, some social researchers try to "give voice" to their research subjects—providing their subjects the opportunity to have their stories told, their worlds represented. If not for the interest or concern of social researchers, these groups might have little opportunity to relate their lives, in their own words, to the literate public. For example, the experiences of recent immigrants struggling for survival in the noise and confusion of our largest and most congested cities are rarely represented in the media.

The goal of giving voice clearly does not follow from the model of the hard sciences. A physicist is usually not concerned to give voice to the lives and subjective experiences of specific particles. The goal of giving voice may come into direct conflict with the goal of identifying general patterns because it is difficult to both privilege certain cases by giving them voice and at the same time chart general patterns across many cases. When the goal is to identify general patterns, no specific case, no specific voice, should dominate.

Altogether, seven major goals of social research are examined in this chapter. They include:

1. identifying general patterns and relationships
2. testing and refining theories
3. making predictions

4. interpreting **culturally or historically significant phenomena**
5. exploring diversity
6. giving voice
7. advancing new theories

Generally, the first three goals follow the lead of the hard sciences. The fourth and sixth goals, by contrast, follow from the social nature of social science—the fact that social researchers study phenomena that are relevant in some special way to the social world of the researcher. The fifth and seventh goals straddle these two domains. In some ways they link up with hard science models; in other ways they reflect the socially grounded nature of social research.

The list of goals discussed in this chapter is not exhaustive; several others could be added. For example, **evaluation research,** which is a type of social research, seeks to measure the success of specific programs or policies, especially in education and social services. Did the clients of an agency benefit when its record keeping procedures were simplified and streamlined? Or did the resulting sacrifice of detailed information following the effort to streamline harm specific categories of clients? Which ones? While evaluation research usually has very specific goals tied to particular programs, such research is also relevant to general patterns, one of the key concerns of social research. Thus, most social research involves at least one and usually several of the seven goals discussed in this chapter.

Because social research has multiple and competing goals, a variety of different research strategies have evolved to accommodate different goals. A **research strategy** is best understood as the pairing of a primary *research objective* and a specific *research method.* The last part of this chapter introduces three common research strategies, among the many different strategies that social researchers use. The three research strategies discussed in this chapter and examined in detail in Part II of this book are

1. *qualitative research* on the commonalities that exist across a relatively small number of cases
2. *comparative research* on the diversity that exists across a moderate number of cases
3. *quantitative research* on the correspondence between two or more attributes across a large number of cases (covariation)

The Seven Main Goals

1. *Identifying General Patterns and Relationships*

Recall that one of the key characteristics of social scientific representations discussed in Chapter 1 was their focus on social phenomena that are socially significant in some way. Phenomena may be significant because they are common or *general;* they affect many people, either directly or indirectly. This quality of generality makes knowledge of such phenomena valuable. For example, suppose it can be shown that in countries where more public funds are spent on the prevention of illness (for example, by improving nutrition, restricting the consumption of alcohol and tobacco, providing children free immunization, and so on), health care costs less in the long run. Knowledge of this general pattern is valuable because it concerns almost everyone.

One of the major goals of social research is to identify general patterns and relationships. In some corners, this objective is considered the *primary* goal because social research that is directed toward this end resembles research in the hard sciences. This resemblance gives social research more legitimacy, making it seem more like social physics and less like social philosophy or political ideology.

For most of its history, social research has tried to follow the lead of the hard sciences in the development of its basic research strategies and practices. These approaches to research are especially well suited for examining general patterns, and knowledge of general patterns is a highly valued form of knowledge. For example, if we know the general causes of ethnic antagonism (one general cause might be the concentration of members of an ethnic minority in lower social classes), we can work to remove these conditions from our society or at least counteract their impact and perhaps purge ourselves of serious ethnic antagonism. As more and more is learned about general patterns, the general stock of social scientific knowledge increases, and it becomes possible for social scientists to systematize knowledge and make connections that might otherwise not be made. For example, general knowledge about the causes of ethnic antagonism within societies might help to further understanding of nationalism and the international conflicts spawned by national sentiments.

Knowledge of general patterns is often preferred to knowledge of specific situations because every situation is unique in some way. Understanding a single situation thoroughly might be pointless if this understanding does not offer *generalizable* knowledge—if it doesn't lead to some insight relevant to other situations. From this perspective, knowing one situation thoroughly might even be considered counterproductive because we could be deceived into thinking an atypical situation offers useful general knowledge when it does not, especially if we are ignorant of how this situation is atypical.

Because of the general underdeveloped state of social scientific knowledge, we are not always sure which situations are typical and which are not. Furthermore, because every situation is unique in some way, it also could be argued that every situation is atypical and therefore untrustworthy as a guide to general knowledge. In short, when the goal is knowledge of general patterns, social researchers tend to distrust what can be learned from one or a small number of cases.

According to this reasoning, knowledge of general patterns is best achieved through examination of many comparable situations or cases, the more the better. The examination of many cases provides a way to neutralize each case's uniqueness in the attempt to grasp as many cases as possible. If a broad pattern holds across many cases, then it may reflect the operation of an underlying cause which can be inferred from the broad pattern. (On issues of plausible inference, see Polya 1968.)

For example, while it may be possible to identify both "kind and benevolent" dictators and democratic governments that terrorize their own citizens, the broad pattern across many countries is that the more democratic governments tend to brutalize their own citizens less. This correspondence between undemocratic rule and brutality, in turn, may reflect the operation of an underlying cause—the effect that the concentration of power has on the incidence of brutality. While not directly observed, this cause might be inferred from the observed correspondence between undemocratic rule and brutality. It is obvious that brutality and benevolence exist in all countries. Still, across many cases the pattern is clear, and exceptions should not blind us to the existence of patterns.

2. Testing and Refining Theories

General patterns matter not only because they affect many people, but also because they are especially relevant to social theory. As described in Chapter 1, social theories come out of a huge, on-going conversation among social scientists and other social thinkers. This conversation is an ever-changing pool of ideas, a resource to draw on and to replenish with fresh thinking.

It is also important to note that there is a virtually limitless potential for new ideas to emerge from within this pool because existing ideas can be combined with each other to produce new ones, and new implications can be drawn from these new combinations. Also, social theory is forever borrowing ideas from other pools of thinking, including philosophy, psychology, biology, and even physics, chemistry, and astronomy. The cross-fertilization of ideas is never ending.

For example, ideas about the relationship between workers and owners in industrial countries, especially the idea that workers are exploited, have been applied to the relations between countries. Some analyses of work emphasize the degree to which profits are based on keeping the wages of workers low, especially those with the fewest skills. From this perspective, there is natural conflict between the owners of firms and the workers: If wages are kept low, then profits will be higher; if wages are too high, profits will suffer.

This thinking has been transferred to the international arena by some theorists who assert that rich countries benefit from the poverty of poor countries (see, for example, Baran 1957; Frank 1967, 1969; Wallerstein 1974, 1979). Some theorists argue that labor-intensive production, which uses simpler technologies and tends to offer only very low wages, has been shifted to poor countries, while the rich countries have retained capital-intensive production, which uses advanced technology. Workers in rich countries benefit from the greater availability of high-wage jobs and from the cheap prices of the labor-intensive goods imported from low-wage countries. In this way, all the residents of rich countries—owners, managers, *and* workers—exploit the cheap labor of poor countries (see Lenin 1975).

This argument, which is an example of the cross-fertilization of ideas, can be tested with economic data on countries. In this way, a new perspective—and a new source for testable hypotheses—is derived from existing ideas.

One of the primary goals of social research is to improve and expand the pool of ideas known as social theory by testing their implications, as in the example just presented, and to refine their power to explain. Typically, this testing is done according to the general plan of the scientific method, as described in Chapter 1. Hypotheses are derived from theories and their implications and then tested with data that bear directly on the hypotheses. Often the data are collected specifically for testing a particular hypothesis, but sometimes already existing data can be used (for example, census and other official statistics published by government agencies).

By testing hypotheses, it is possible to improve the overall quality of the pool of ideas. Ideas that fail to receive support gradually lose their appeal, while those that are supported more consistently gain greater stature in the pool. While a single unsuccessful hypothesis rarely kills a theory, over time unsupported ideas fade from current thinking. It is important to identify the most fertile and powerful ways of thinking and to assess different ideas, comparing them as explanations of general patterns and features of social life. Testing theories can also serve to refine them. By working through the implications of a theory and then testing this refinement, it is possible to progressively improve and elaborate a set of ideas.

It is possible to conduct social research without paying much direct attention to this pool of ideas. There are many aspects of social life and many different social worlds that attract the attention of social researchers, independent of the relevance of these phenomena to social theory. After all, social researchers, like most social beings, are curious about social life. However, improving the quality of social theory is an important goal because this pool of ideas structures much thinking and much telling about society, by social scientists and by others.

3. *Making Predictions*

While social researchers use theories to derive "predictions" (actually, hypotheses) about what they expect to find in a set of data (for example, a survey), they also use accumulated social scientific knowledge to make predictions about the future and other novel situations. It is this second meaning of the word **prediction** that is intended when we say that "making predictions" is one of the major goals of social research.

Consider an example of this second kind of prediction: Research indicates that ethnic conflict tends to increase when the supply of economic rewards and resources (jobs and promotions, for instance) decreases. Thus, a social scientist would predict increased ethnic tensions in an ethnically diverse country that has just experienced a serious economic downturn. Prediction is often considered the highest goal of science. We accumulate knowledge so that we can anticipate things to come. We make predictions based on what we know. Two kinds of knowledge help us make predictions. Knowledge of history (past successes and failures) and knowledge of general patterns.

Knowledge of history helps us to avoid repeating mistakes. Understanding of the Stock Market Crash of 1929 and the ensuing Great Depression, for example, has motivated our economic and political elites to moderate the violent swings of market-oriented economic life. An unsuccessful military venture into Vietnam in the 1960s and 1970s has made our military leaders wary of intervening in guerrilla wars. Social researchers draw lessons from history by relating events to general concepts. The Stock Market Crash of 1929 provides clear lessons about the need that arises for a balance between the free play of markets (for example, stock markets) and regulations imposed through hierarchies (for example, the Securities Exchange Commission). The prediction here is that unregulated markets will fluctuate widely and may even self-destruct.

The second kind of knowledge, understanding of general patterns, is useful for making projections about likely future events. For example, we know that certain types of crime (drug dealing, for instance) increase when legitimate economic opportunities decrease. We can use this knowledge combined with assumptions about other causal factors to extrapolate future crime rates given different employment conditions. If current trends toward higher production levels with fewer workers continue, it would seem reasonable to anticipate increases in certain types of crimes. Projections of this type are quite common and sometimes can be surprisingly accurate. It is much easier to predict a rate (the rate of homelessness, the rate of drug-related crimes, the rate of teenage pregnancy, and so on) than it is to predict what any single individual might do. For example, it is easy to extrapolate or project a good estimate of the number of people who will be murdered in Los Angeles next year, but it is impossible to predict very much about who, among the millions, will be the perpetrators or the victims.

While making predictions is one of the most important goals of social research, it's not always the case that prediction and understanding go hand in hand. Sometimes our predictions are quite accurate, but our understanding of the actual underlying processes that produce outcomes is incomplete or simply erroneous. For example, the causes of drug addiction are quite complex, as is the process of becoming an addict. However, it is a relatively simple matter to forecast levels of drug addiction in major U.S. cities based on knowledge of the social conditions that tend to favor high levels of addiction.

A simpler example: It might be possible to predict with fair precision how many murders will be committed next year based on the number of automobiles stolen this year. However, that doesn't mean that some fixed percentage of the people who steal cars one year graduate to homicide the next. More than likely, the two rates both respond to the same causal conditions (such as unemployment or the formation of street gangs), but at different speeds.

Predicting rates is much easier than predicting specific events. The kinds of things many social scientists would like to be able to predict—namely, the occurrence of specific events at specific points in time in the future—are simply beyond the scope of any science. For example, many social scientists chastised themselves for being unable to predict the fall of communism in Eastern Europe in 1989. Their failure to predict these dramatic events made them feel impotent. However, no science, social or otherwise, could possibly achieve this kind of prediction—the timing of specific future social or natural events. The key to understanding this is the simple fact that it is very difficult to predict *specific* future events.

Consider the "hard" science of meteorology. At best, this science can predict the probability of rain over the next several days. But what if we want to know when it will start, when it will stop, and how much it will rain? It should be possible to predict these things. After all, no human intervention, interpretation, or subjectivity is involved, only measurable, physical qualities like temperature, wind direction and velocity, moisture, and so on. But the hard science of meteorology cannot offer this precision; it simply cannot predict specific events. Nor can meteorology predict which day, or even which year, a hurricane will again sweep across Galveston Island, Texas. Even when there is a hurricane in the middle of the Gulf of Mexico, it's very difficult to tell which, if any, coastal area it will demolish.

In a similar manner, no social scientists could predict, say in 1980, that communism would fall in Eastern Europe *in 1989*. For many years, some social scientists claimed that communism was likely to fall in the near future. Even in 1980 a few would have been willing to attach specific probabilities to specific years, say a 40% chance of falling by the year 2000. Social science is not impotent, but appears so because of the specificity of the predictions we desire.

Will a new religious movement, emphasizing conservative values, the sanctity of marriage and the family, self-reliance, and the rejection of white culture and its materialism sweep black inner-city neighborhoods next year? Sometime in the next ten years? Will wild spasms of nihilistic self-destructiveness sweep through teenage populations in the white suburbs of major U.S. cities in the year 2009? It would certainly be impressive to be able to predict events such as these, but it is outside the scope of any science to offer this degree of specificity. At best, social researchers can make broad projections of possibilities using their knowledge of general patterns.

4. Interpreting Culturally or Historically Significant Phenomena

Knowledge of general patterns is not the only kind of valuable knowledge, however, especially when it comes to understanding social life. In the social sciences, knowledge of specific situations and events, even if they are atypical (and usually *because* they are atypical; see Dumont

1970), is also highly valued. The significance of most historical phenomena derives from their atypicality, the fact that they are dramatically nonroutine, and from their impact on who we are today.

For example, many social researchers address important historical events like the French Revolution or the civil rights movement. We care about these events and their interpretation (for example, how the Roman Empire fell or the history of slavery) because of the relevance of these events for understanding our current situation—how we got to where we are. We are fascinated by the U.S. Civil War not because we expect it to be repeated, but because of its powerful impact on current race relations and the structure of power (who dominates whom and how they do it) in the United States today.

Other phenomena are studied not because of their historical relevance to current society but because of their cultural relevance. The bits and pieces of African cultures that slaves brought with them, for example, have had a powerful impact on the course and development of American culture. Other phenomena may be culturally significant because of what they may portend. The heavy metal rock culture of the late twentieth century, for example, could signal future directions of American culture.

Often there is competition among social researchers to establish the "accepted" interpretation of significant historical or cultural phenomena. For example, social researchers have examined the events that led to the fall of communist regimes (that is, of the power cliques that controlled the centrally planned economies of Eastern Europe). These events have been addressed because they are historically and culturally relevant and significant, and different researchers have different ideas about how and why these regimes fell. The interpretation of these events that prevails, especially the interpretation of the fall of the communist regime in the former Soviet Union, has important implications for how both social scientists and the literate public think about "communism" and the possibility of centralized control of national economies. It is not always the case that a single interpretation prevails, not even in the very long run. The struggle to have an interpretation accepted as "correct" can extend over generations of scholarship and stretch over centuries of debate.

Social researchers who study general phenomena usually do not address specific events or their interpretation. They would rather know about a general pattern (for example, the covariation across countries between the extent to which democratic procedures are practiced, on the one hand, and the level of political repression, on the other) than about a specific set of events (the detention of Japanese-Americans by the U.S.

government during World War II, for instance). It is difficult, however, to address many of the things that interest social researchers, and their audiences, in research that focuses only on what is general.

For example, social researchers sometimes address the subjectivity or consciousness of their subjects. There are many possible interpretations for any set of events. Did the Nazis intend to exterminate the Jews all along, or did they adopt this policy in response to the conditions of World War II? Was it necessary for Stalin to terrorize Soviet citizens in order to forge state socialism? Was he insecure and paranoid, or was terrorism simply an effective way of maintaining his personal power? In both episodes of massive inhumanity, it is not enough to know that millions of people died or how they died. We want to know why. Researchers who study general patterns typically do not address issues related to the consciousness of their research subjects.

5. Exploring Diversity

Another major goal of social research is to explore and comprehend the social diversity that surrounds us. While this goal may seem similar to the goal of identifying general patterns, and does complement it in some respects, it is quite different. For example, one general pattern is that educational and economic development tend to go together; countries with better schools and higher literacy rates tend to be richer. However, the fact that a general pattern exists doesn't mean that there aren't important and interesting exceptions. Some poor countries have well-developed educational systems and very high literacy rates (for example, Sri Lanka), and some rich countries have poorly developed schools and surprisingly low levels of literacy (Saudi Arabia, for instance).

Exploring diversity often means that the researcher ignores dominant patterns and focuses on the *variety* of circumstances that exist. How is living in a poor country with a high level of literacy different from living in other poor countries? What happens when a low level of educational development or literacy is combined with wealth? In short, the study of diversity avoids an exclusive focus on what is most common or on dominant patterns.

More generally, exploring diversity furthers an understanding and appreciation of **sociodiversity**, a concept that parallels the ecological notion of biodiversity. We protect biological species close to extinction because we are concerned about biodiversity. The human species dominates all others, so much so that many species are threatened with extinction. Many environmentalists see declining biodiversity as an indicator of the degree to which human societies have threatened the self-regulating natural order of the biosphere we call Earth.

People are less concerned about sociodiversity. Anthropologists have documented dramatic declines in sociodiversity. They have studied societies in all corners of the world over much of the last century. As the reach of global economic and political forces has expanded, these forces have more deeply penetrated many parts of the world. Small-scale societies that were once more or less external to the international system have been incorporated into it. One direct consequence of this incorporation is the disappearance of many cultural forms and practices and the transmutation of countless others. Sociodiversity at the level of whole societies has declined dramatically. More and more, there is a single, dominant global culture.

A simple example of this change is the worldwide decline in arranged marriages and the increased importance of romantic involvement (see Barash and Scourby 1970). From the perspective of modern-day U.S. Americans, this shift seems natural and inevitable, and arranged marriages seem quaint. But in fact arranged marriages have been an important source of social order and stability in many societies, joining different families together in ways that undercut social conflict.

The efforts of anthropologists to document rapidly disappearing societies have been preserved in their writings and in data compilations such as the **Human Relations Area File** (HRAF), which catalogs many different aspects of hundreds of societies and cultures that no longer exist. It is important to understand societies that differ from our own because they show alternative ways of addressing common social issues and questions. For example, societies cope with scarcity in different ways. In some societies great feasts involving entire communities are a routine part of social life. These feasts not only provide protection against starvation, especially during lean years, but they also increase the strength of the social bonds joining members of communities. There has also been remarkable diversity among human societies in how basic arrangements like the family, kinship, the gender division of labor, and sexuality have been structured or accomplished.

Of course, great social diversity exists today, despite the impact of that giant steamroller, the world capitalist economy, on sociodiversity worldwide. There are many social worlds (and social worlds within social worlds—see Chapter 1) in all parts of all countries. There is great diversity even in the most advanced countries—those most closely joined by the world economy. Often, much diversity is simply unacknowledged or ignored. Sometimes assumptions are made about sameness (for example, that people living in inner-city ghettos think or act in certain ways) that turn out to be false when the diversity within a social category is examined closely. Also, people often respond to sameness and uniformity by crafting new ways of differentiating themselves from others. Sometimes, these efforts lead only to new fads; sometimes, they culminate in entirely new social formations (as when a religious cult withdraws from mainstream society).

Sometimes social researchers start out not knowing if studying a new case or situation will offer useful knowledge of diversity. They study it in order to make this assessment. For example, some immigrant groups are very successful. It is important to find out how and why they are successful in order to determine if this knowledge is relevant to other groups (or, more generally, to U.S. immigration policy). It may be that their success is due to circumstances that cannot be duplicated elsewhere. But there is no way to know this without studying the specific causes of their success. Another example: Catholic nuns tend to live longer and healthier lives than most other groups, religious or secular. It may not be the case that we have to live like nuns to match their longevity, but we won't know this unless we study them and find out why they live longer than others. Whether or not the study of diverse groups offers knowledge that is useful, research on diverse groups contributes to social scientists' understanding of social life in general.

6. Giving Voice

Sometimes the goal of exploring diversity is taken one step further, and the researcher studies a group not simply to learn more about it, but also to contribute to its having an expressed voice in society. In research of this type, the objective is not only to increase the stock of knowledge about different types, forms, and processes of social life, but to tell the story of a specific group, usually in a way that enhances its visibility in society.

Very often the groups studied in this way are marginal groups, outside the social mainstream (for example, the homeless, the poor, minority groups, immigrant groups, homosexuals, people labeled mentally ill, and so on). This approach to social research asserts that every group in society has a "story to tell." Some groups (for example, business people, middle class whites, and so on), are presented in the mainstream beliefs and values of society as the way life is and should be. Many social researchers believe that it is their responsibility to identify excluded groups and tell their stories. By giving voice, researchers often are able to show that groups that are considered deviant or different in some way do not deviate as much as most people think. For example, a common finding is that even people in the most dire and difficult circumstances strive for dignity.

While social researchers who do this kind of research often focus on marginal or deviant groups, this emphasis is neither necessary nor universal. Arlene Daniels (1988), for example, studied the wives of rich and powerful men in a West Coast city and argued that many of them carried on what she called "invisible careers." In a book bearing that title she documented their tireless charitable activities and showed how these privileged women organize volunteer efforts to improve the quality of life in their communities. Still, their efforts are hidden and taken for granted, and the women themselves portray their labor not as work but as self-sacrifice.

In research that seeks to give voice, social theories may help the researcher identify groups without voice and may help explain why these groups lack voice, but theory is not considered a source of hypotheses to be tested. When the goal of a project is to give voice to research subjects, it is important for the researcher to try to see their world through their eyes, to understand their social worlds as they do. Thus, researchers may have to relinquish or "unlearn" a lot of what they know in order to construct valid representations of their research subjects—representations that embody their subjects' voice.

To achieve this level of in-depth understanding, researchers must gain access to the everyday world of the group. It might be necessary, for example, to live with the members of a marginalized group for extended periods of time and gradually win their confidence (see, for example, Stack 1974; Harper 1982). When the researcher feels he or she knows enough to tell their stories, one goal of the telling might be to try to minimize, as much as possible, the voice of the researcher.

Some researchers, for example, take photographs of the social worlds of a group and then record their subjects' descriptions and interpretations of the photographs. A transcript of their descriptions is then published along with the photographs (see Harper 1987; Suchar and Markin 1990). In fact, a variety of systematic techniques have been developed by social researchers to facilitate this type of in-depth knowledge and understanding (see Denzin 1970, 1978; McCall and Simmons 1969; Strauss 1987)

Some social researchers consider research that seeks to give voice *advocacy* research and therefore doubt its objectivity. (Becker 1967 addresses this issue in depth.) How can research that seeks to enhance the visibility of a marginal group be conducted in a neutral way? Isn't it inevitable that researchers will favor the positive aspects of marginal groups in their representations of these groups? Most social researchers are committed to objectivity and neutrality in much the same way that most journalists are. Some common cautions are

don't whitewash

present the good and the bad

be wary of how people rationalize what they do

maintain skepticism

examine the same events from several points of view

Giving voice does not necessarily entail advocacy. Still, social researchers who seek to give voice must be vigilant in their efforts to represent their groups appropriately. Most social worlds, marginal or mainstream, are quite complex. Advocacy typically oversimplifies. Generally, it is not difficult to spot a one-sided representation or to recognize research that merely advocates for a group.

Those who argue that giving voice is not a valid research objective should acknowledge that almost all research gives voice in the sense that it enhances the visibility of the thing studied and represents the viewpoint of some group or groups, even implicitly. Even a study of the general social conditions that favor stable democracy across many countries enhances the importance and visibility of stable democracy as a desirable condition simply by studying it. Research that seeks to give voice is clear in its objectives.

7. Advancing New Theories

Many different kinds of social research advance social theory, even research that seeks to interpret historical or cultural significance. The testing of theories (goal 2) also advances theory in the limited sense that these tests indicate which theoretical ideas have more support as explanations of social life. The goal of advancing theory as it is used here, however, involves more than assessing and refining existing ideas. When theory is advanced, ideas are elaborated in some *new* way. To advance theory it is not necessary to come up with a complete model of society or even some part of it. The development of new ideas and new concepts is the most that research seeking to advance theory usually accomplishes.

Theory testing (goal 2) is primarily *deductive*. Hypotheses about social life are derived from theories and then tested with relevant data. The researcher then draws the implications of the results of these tests for theory (see Chapter 1). Research that advances theory, by contrast, is usually described as having an *inductive* quality. On the basis of new evidence, the researcher develops a new theoretical concept or new relationship or advances understanding of existing ones.

Not only does the researcher use data to illustrate the new concept, he or she may also elucidate the relation of the new concept to existing concepts. One researcher, for example, developed the concept of *edgework* based on his studies of people who skydive and from related research on people who seek out other dangerous situations (Lyng 1990). When developing a new concept, it is necessary to distinguish it from related concepts and to explain its logical and causal connections to others (see also Wieviorka 1988, 1992).

Many theoretical advances come from detailed, in-depth examination of cases. Exploring diversity, for example, may lead to the discovery of new social arrangements and practices. The study of behavior of the groupies who surround certain kinds of rock bands, for example, might lead to new insights about the importance of rituals in contemporary social life. The mere existence of novel phenomena also may challenge conventional thinking. Existing theories may argue that certain ways of doing things or certain behaviors are incompatible, that it has to be either one or the other. The discovery that "incompatible" elements can coexist calls such theories into question and may force researchers to theorize about how such logically incompatible things can coexist.

Research that gives voice also may lead to theoretical advances because such research often leaves existing theories behind in its attempt to see social worlds through the eyes of their members. This openness to the viewpoints of low-status and low-visibility people may expose the inadequacies of existing theoretical perspectives. Finally, work that seeks to interpret cultural or historical significance may also advance theory because it too is based on detailed analyses of cases. For example, in-depth research on the Iranian revolution could lead to new insights on the importance of the interplay of religious ideology and political organization in large-scale political change.

Research that seeks to identify general patterns across many cases is usually associated with the goal of testing theory (via hypotheses), and less often with the goal of advancing theory, even though, as already noted, testing theory does refine it. However, the analysis of broad patterns can lead to theoretical advances (see, for example, Paige 1975; Rokkan 1970, 1975; Tilly 1984; Rueschemeyer et al. 1992). Sometimes hypotheses fail or are only partially supported, and researchers generally want to know why. They may study additional patterns in their data to find out why the theory they are testing does not fit the data well.

For example, using a generally accepted theory as a starting point, a researcher might test the hypothesis that richer countries tend to have a more equal distribution of income (that is, within their own borders) than poorer countries. Analysis of relevant data might show that while this pattern holds for most countries, among the richest fifteen or so it does not—they might all have roughly the same degree of equality. This finding might lead the researcher to speculate about the newly discovered pattern: Why is it that greater national wealth does not lead to greater equality once a certain level of economic development is reached? A variety of factors might be examined in the effort to account for this pattern. This search might lead to the identification of causal factors that suggest fundamental revision of the theory used to generate the initial hypothesis about patterns of income inequality.

While the deduction-versus-induction distinction is a simple and appealing way to differentiate kinds of social research, most research includes elements of both (see Stinchcombe 1968). For this reason some philosophers of science (for example, Hanson 1958) argue that all research involves **retroduction**—the interplay of induction and deduction. It is impossible to do research without some initial ideas, even if the goal is to give voice to research subjects. Thus, almost all research has at least an element of deduction. Similarly, almost all research can be used to advance theory in some way. After all, social theories are vague and imprecise. Every test of a theory refines it, whether or not the test is supportive. Research involves retroduction because there is typically a dialogue of ideas and evidence in social research. The interaction of ideas and evidence culminates in theoretically based descriptions of social life (that is, in social scientific representations) and in evidence-based elaborations of social theory.

The Link between Goals and Strategies

It is clear that no researcher can tackle all seven goals at once, at least not in the same study. A classic view of science says that it is a violation of the scientific method to try to advance theory (goal 7) and test theory (goal 2) in the same study. Data used to generate a new theory should not also be used to test it. Most of the tensions between goals, however, revolve around practical issues.

It is difficult, for example, to both examine *many* cases so that a general pattern can be identified (goal 1) and study *one* case in depth so that its specific character can be understood (goal 6). Even when it is possible to do both, they don't always mix well. What if the findings from the in-depth study of one or a small number of cases contradict the results of the analysis of broad patterns across many cases? Which finding should the social researcher trust? However, both kinds of research are important because both help social researchers find order in complexity, order that they can represent in their reports. The first type of research helps social researchers identify what is general across many cases—to discern the underlying order that exists amid great variation; the other helps them comprehend the complexity of specific situations directly.

Many different strategies of social research have emerged to accommodate its multiple and competing goals. As already noted, a research strategy is best understood as a pairing of a general research objective and a specific research method. Each strategy constitutes a way of linking ideas and evidence to produce a representation of some aspect of social life. Research strategies structure how social researchers collect data and make sense of what they collect. Even though some strategies are clearly more popular than others, there is no single "correct" way of conducting social research.

While there are many different strategies of social research, three very broad approaches are emphasized here:

- the use of qualitative methods to study commonalities
- the use of comparative methods to study diversity
- the use of quantitative methods to study relationships among variables

These three strategies are discussed in detail in Part II of this book because they represent three very common but very different ways of carrying on a dialogue of ideas and evidence. The selection of these three strategies does not imply that other strategies are not important or do not exist. Indeed, there are plenty of qualitative researchers who study diversity, and there are many researchers who use comparative methods to study commonalities. The pairings emphasized here (qualitative methods with commonalities, comparative methods with diversity, and quantitative methods with covariation) have been selected because they offer the best illustration of the core features of different methods. They also provide strong testimony to the unity and diversity of social research.

Qualitative researchers interested in commonalities examine many aspects or features of a relatively small number of cases in depth. A study of how one becomes a marijuana user (Becker 1953) is an example of a qualitative study.

Comparative researchers interested in diversity study a moderate number of cases in a comprehensive manner, though in not as much detail as in most qualitative research. A study of the checkered history of democratic institutions in South American countries is an example of a comparative study (E. Stephens 1989).

Quantitative researchers interested in how variables covary across cases typically examine a relatively small number of features of cases (that is, variables) across many, many cases. A study of the correspondence between the intensity of party competition and the level of voter turnout across all counties in the United States is an example of a quantitative study.

These three strategies can be plotted in two dimensions showing the relation between the number of cases studied and the number of aspects of cases studied (see Figure 2.1). The figure illustrates the trade-off between studying cases and studying aspects of cases, or variables. Because the energies and capacities of researchers are limited, they often must

FIGURE 2.1

Cases, Aspects of Cases, and Research Strategies*

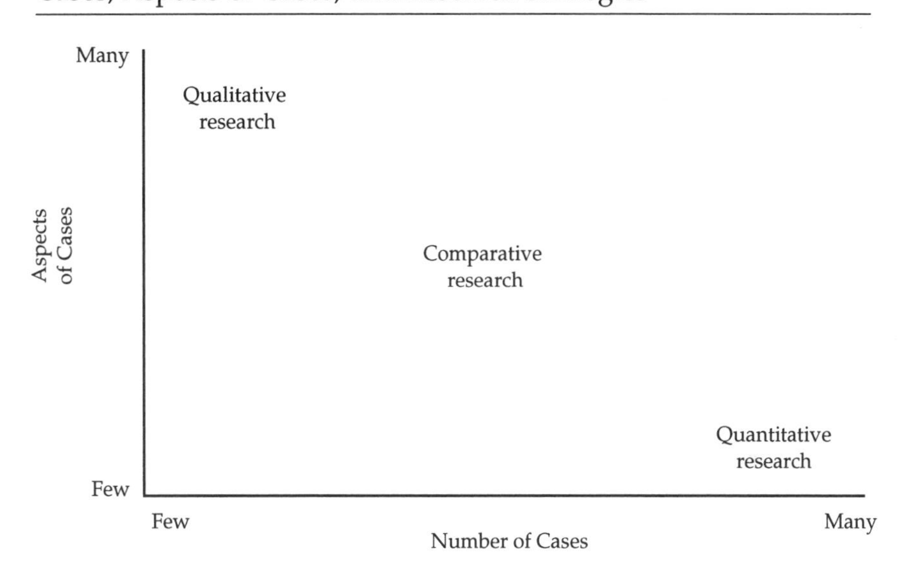

*The three research strategies are qualitative research on commonalities, comparative research on diversity, and quantitative research on relationships between variables.

choose between focusing on cases as wholes (qualitative research on commonalities), focusing on variables (quantitative research on relationships among variables), or balancing the two in some way (comparative research on diversity). It is possible to gain a detailed, in-depth knowledge of a small number of cases, to learn a moderate amount about an intermediate number of cases, or to focus on limited information from a large number of cases.

The trade-off between number of cases and number of features does not concern how much information social researchers can *collect*. After all, social researchers can collect volumes of information on each of thousands and thousands of cases (Davis and Smith 1988). The Internal Revenue Service collects detailed information on millions of people every year. The issue is how much information social researchers, or anyone else for that matter, can *study; how* the information is studied (for example, is each case examined individually?); and the *relevance* of the information to a particular research question.

Imagine trying to grasp the nature of informal interpersonal networks in each of the top 500 U.S. corporations. It might take years to unravel the informal networks of a single corporation. A social researcher can gain this kind of intimate knowledge about only a relatively small number of cases.

However, it might be possible to survey these same 500 corporations and find out basic information like total assets, profitability, and number of employees. The information from this survey would not add up to intimate knowledge of each of the 500 corporations, but could be used to examine relations among variables characterizing these corporations. For example, does large corporate size pose an obstacle to profitability? Answering this question does not require in-depth knowledge of the workings of *any* of the 500 corporations. Of course, such in-depth knowledge would improve the analysis of the evidence on size and profitability and the representation of the results, but it is not essential to the study of the general relationship between these two variables.

It is important to note that Figure 2.1 represents the tendencies of these three strategies and does not establish absolute boundaries around these three strategies in any way. Some quantitative researchers, for example, collect hundreds of variables on thousands of cases when they conduct research, and they try to squeeze as much of this information as possible into the representations they construct. Of course, these representations are still "big picture" representations of broad patterns of covariation across cases. Likewise, there are some qualitative researchers who work in teams to increase the number of cases they study. Thus,

Figure 2.1 should be viewed as an attempt to depict the nature of the typical representations that result from these three common strategies.

Table 2.1 maps the relation between these three strategies and the seven goals of social research discussed in this chapter. The column headings of the table are the three general strategies; the rows are the seven goals. The table shows the fit between goals and strategies.

The three different strategies range from intensive (qualitative study of commonalities) to comprehensive (comparative study of diversity) to extensive (quantitative study of the relationships among variables) in their approach to cases. An intensive approach is best suited for goals that involve close attention to specific cases; a comprehensive approach is best suited for goals that involve examination of patterns of similarities and differences across a moderate number of cases; an extensive approach is best suited for goals that involve knowledge of broad patterns across many cases. It is important to remember, however, that the three strategies examined here and in Part II are three among many different strategies of social research.

The goal of identifying general patterns (goal 1), for example, is best served by the quantitative approach, but it is also served by the comparative approach, though maybe not quite as well. (Thus, the *primary* strategy for identifying general patterns is the quantitative approach; a *secondary* strategy is the comparative approach.) A pattern is not general if it does not embrace many cases. Also, most statements about general

TABLE 2.1

The Goals and Strategies of Social Research*

	Qualitative Research	*Comparative Research*	*Quantitative Research*
1. Identifying broad patterns		secondary	primary
2. Testing/refining theory	secondary	secondary	primary
3. Making predictions		secondary	primary
4. Interpreting significance	primary	secondary	
5. Exploring diversity	secondary	primary	secondary
6. Giving voice	primary		
7. Advancing new theories	primary	primary	secondary

*The three research strategies are qualitative research on commonalities, comparative research on diversity, and quantitative research on relationships between variables. *Primary* indicates that the strategy is a very common way of achieving a goal; *secondary* indicates that the strategy is sometimes used to achieve a goal.

patterns involve variables. Both of these features of general patterns point to the quantitative approach as the primary strategy. The goal of testing theory (goal 2) is served by all three strategies. Most theories, however, are composed of abstract concepts that are linked to each other and thus concern general relationships that can be viewed across many cases or across a range of cases. Sometimes a single case will offer a critical test of a theory, but this use of individual cases is relatively rare (Eckstein 1975). Besides, from the perspective of most theories, single cases are unique and therefore relatively unreliable as raw material for testing theories. Likewise, the most appropriate strategy for making predictions is the quantitative approach. Most predictions involve extrapolations based on many cases, the more the better, as long as they are appropriate and relevant to the substance of the prediction.

The goals of interpreting significance and giving voice, by contrast, are best served by a strategy that examines a small number of cases (often a single historical episode or a single group) in depth—the qualitative approach. Similarly, the best raw material for advancing theory is often provided by strategies that focus on cases, which is the special forte of qualitative research and one of the strong points of comparative research. However, all research, including quantitative research, can advance theory. Finally, the goal of exploring diversity is best served by the comparative approach. However, because qualitative and quantitative research contribute to knowledge of diverse groups, they too serve this goal.

The Social Nature of Social Research

Imagine a chart comparable to Table 2.1 constructed for a hard science like chemistry or physics. Goals 4 and 6 would not exist, at least they would not be considered major goals, and goal 5 would concern only a handful of researchers. The remaining four goals (1, 2, 3, and 7) are all served by the quantitative approach—a strategy that addresses general relations between measurable aspects of the things social scientists study. Goals 4, 5, and 6 reflect the social nature of social research. It is also these goals that sometimes make social scientists seem "unscientific," especially to scientists, social or otherwise, strongly committed to the other goals.

Consider again the goal of giving voice. Why should any particular voice be privileged by social research? Why should a social researcher try to enhance a particular group's visibility in society? Who cares whether or not people who are not marginal can understand those who are? Consider the goal of interpreting cultural or historical significance.

How do we know that the social researcher is not trying to whitewash horrific events, or perhaps make the members of a truly destructive group look like victims of oppression? Finally, consider the goal of exploring diversity. By highlighting diversity, a social researcher may glorify it. But too much diversity in society can tear it apart. Might it be better to emphasize the things that we have in common, what most members of society share?

These aspects of social research make it an easy target of criticism. However, it is important to understand that no social research exists in a vacuum. Research on general patterns, for example, may simply privilege what is normative. All social research gives voice in one way or another to some aspect of society. Similarly, research that tests theories has implications for how we think about human nature, social organization, and the different kinds of social worlds that are possible to construct. In fact, because of the social nature of social research, all social research has implications for the interpretation and understanding of anything that people do or refuse to do together. Social research is inescapably social in its implications. For this reason, social researchers cannot escape bias, regardless of which goals motivate research.

3

The Process of Social Research: Ideas and Evidence

Introduction

Social research, in simplest terms, involves a dialogue between **ideas** and **evidence**. Ideas help social researchers make sense of evidence, and researchers use evidence to extend, revise, and test ideas. The end result of this dialogue is a representation of social life—evidence that has been shaped and reshaped by ideas, presented along with the thinking that guided the construction of the representation. This chapter focuses on how the dialogue of ideas and evidence is structured and how it is conducted—how ideas shape the understanding of evidence and how evidence affects ideas.

A major part in the dialogue of ideas and evidence is devoted to the analysis of the phenomena the researcher is studying. The term *phenomena* refers simply to facts or events. **Analysis** means breaking phenomena into their constituent parts and viewing them in relation to the whole they form. A researcher conducting an analysis of a revolutionary movement, for example, might try to dissect it in a way that illuminates all the different forces that combined to make the movement (see Jenkins 1983). This analysis would examine not only the social groups that joined the movement (for example, peasants, workers, soldiers, and so on), but also the social groups that did not; the political and social context, the movement's ideology; and other factors that contributed to its formation.

In essence, the analysis of a revolutionary movement involves breaking it into its key component parts so that it no longer appears to be an amorphous, teeming mass of revolutionaries, but rather can be seen as a combination of key elements and conditions. These elements can be viewed in isolation from one another, and they can be understood in the context of the other parts. For example, the ideology of the movement could be examined both in isolation (What are the key ideas behind the movement?) and in the context of the major groups involved in the movement (How do these key ideas resonate with the concerns of each group—workers, soldiers, peasants, intelligentsia, and so on?). This understanding of the term *analysis*—studying something in terms of its aspects or parts—is necessary background for the concept of *analytic frame,* a key focus of this chapter.

The analysis of social phenomena, while important, is only part of the dialogue of ideas and evidence. The other important part involves the **synthesis** of evidence. Synthesis is the counterpart to analysis. Analysis involves breaking things into parts (for example, the constituent elements of a revolutionary movement); synthesis involves putting pieces together to make sense of them. When social researchers synthesize evidence, they form a coherent whole out of separate parts, making connections among elements that at first glance may seem unrelated. These connections may lead to further insights into the phenomenon they are trying to understand. For example, based on a preliminary examination of evidence from a college sorority, a researcher might develop an initial portrait of it as a type of self-help group. This portrait might be based on interviews with pledges or members. This preliminary synthesis of evidence, in turn, would illuminate other aspects of the sorority which could then be targeted for further study, for example how competition between members is contained.

The process of synthesizing evidence is an important part of the dialogue of ideas and evidence. In this chapter synthesis is presented as a process of forming evidence-based *images* of the research subject. In social research, representations of social life emerge from the interplay between analytic frames (which are derived from ideas) and images (which are derived from evidence).

It is important to examine the different ways the dialogue of ideas and evidence can take shape because the character of the representations of social life that result from different ways of practicing social research are strongly influenced by the nature of this dialogue. For example, the representation of what it is like to be a private in the U.S. Army constructed by a researcher who lives with a group of 5 privates is likely to differ substantially from the representation constructed by a researcher who uses a questionnaire to survey a random sample of 1,000 privates. In both types of research, there is a dialogue of ideas and evidence, but the two dialogues differ dramatically.

This chapter explains how the dialogue of ideas and evidence in social research is carried on through analytic frames (which articulate ideas) and images (evidence-based depictions of social life). The first part of the chapter sketches a simple model of the process of social research as a way to introduce the four basic building blocks of social research:

(1) ideas, (2) analytic frames, (3) evidence, (4) and images. This sketch is presented as a *map* of the ensuing discussion; it is not a full elaboration of the main points of the chapter. Subsequent sections discuss these four building blocks in detail, especially the two that require the greatest clarification, images and analytic frames. The last part of the chapter addresses differences in the interplay of images and analytic frames across three common strategies of social research: the qualitative study of commonalities, the comparative study of diversity, and the quantitative study of covariation.

A Simple Model of Social Research

Figure 3.1 shows the understanding of the process of social research that guides the discussion in this chapter. At the base of the model is evidence. The word *evidence* is the everyday term for what social scientists mean when they use the term *data.* Social researchers use a lot of

FIGURE 3.1

A Simple Model of Social Research

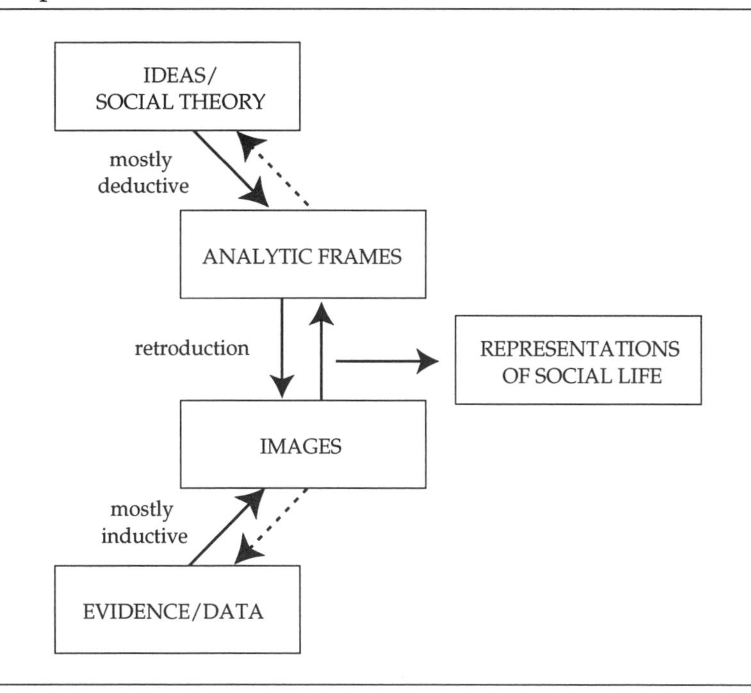

evidence. Studies are often based on the examination of detailed, in-depth information on a small number of cases (as in the qualitative study of commonalities), a moderate amount of information on an intermediate number of cases (as in the comparative study of diversity), or a limited amount of information on many cases (as in the quantitative study of covariation). Ideas are at the top of the model. The word *ideas* is the everyday term for what social scientists call "social theory." Social researchers draw on this pool of ideas when they conduct research, to help them make sense of the things they study.

Ideas and evidence interact through images and analytic frames, shown in the middle of Figure 3.1. Think of an **analytic frame** as a detailed sketch or outline of an idea about some phenomenon. Ideas are elaborated through analytic frames. Frames constitute ways of seeing the things they elaborate.

An analytic frame might be used, for example, to articulate the *idea* of a table. People can recognize a table when they see one, even though tables differ greatly, because they have an *implicit* analytic frame for tables. They understand the category *table* and they can describe how tables vary—in size, color, material used to construct the table, shape of surface, and so on.

The analytic frames of everyday life—like the one for table—are implicit; only rarely are they fully articulated or contested. The analytic frames that guide social research, however, are carefully specified and debated because social researchers must be precise when they define and characterize the phenomena they study. Much of the work of social research centers on debating, clarifying, and using analytic frames to represent social life. These frames make it possible for social researchers to see social phenomena in ways that enhance their relevance to social theory. The analytic frame for revolutionary movements sketched in the introduction to this chapter, for example, provides a brief specification of some of its key components—the different groups involved, their ideologies, and so on.

Images, by contrast, are built up from evidence. Based on observations of workers who run their machines so fast that they break, for example, a researcher might develop an image of these workers as troublemakers or insurgents who subvert production while appearing to work hard. To construct images, researchers synthesize evidence—they connect different parts or elements of the things they study in order to create more complete portraits based on some idea of how these parts are or could be related. Initial images suggest new data collection paths. The researcher working on an image of workers as insurgents who break machines to disrupt work, for example, might study the timing of these disruptions. At what points in the workday, the workweek, or even in the life of a labor contract do these production breakdowns occur? Initial images lead to the collection of more evidence and to a progressive refinement of the image. This image of some workers as insurgents, for example, might lead the researcher to look for other manifestations of subtle subversions of production in this work setting. In short, building images is primarily *inductive*.

This process of synthesizing an image from evidence and refining it goes hand in hand with the process of analyzing the evidence using analytic frames. In essence, by articulating ideas, analytic frames direct an investigation down specific data collection paths. Suppose, for example, in the research just sketched, the researcher had started with an analytic frame for "resistance" that specified a variety of different conditions for its appearance (perhaps using the ideas of Burawoy 1979 or Scott 1976, 1990). This frame might prompt the researcher to consider the subversion of production as a possible form of resistance. The evidence collected, along with other data, might support the image of some workers as insurgents. Once images are built up from evidence they may confirm or amend an analytic frame, or they may summon new ones.

Sometimes the researcher seeks simply to find a good fit between the images constructed from the data and the analytic frames derived from theories. Often though, the fit is not right, and the researcher must determine whether different images can be constructed from the data or whether different analytic frames can be derived from theories. Alternatively, the researcher may use the images constructed from the data to devise new analytic frames or revise old ones. The interaction between analytic frames and images leads both to progressively refined images of social life and to better-specified analytic frames.

This process of refining images culminates in the representation of social life the researcher offers in a report of the results of a study. A social scientific representation thus can be seen as a product of the interaction between images and analytic frames. It is evidence shaped by ideas, which in turn may have been selected and perhaps revised in response to evidence. The subsequent sections of this chapter elaborate the model sketched in Figure 3.1. Of special importance in this discussion are the less familiar notions of images and analytic frames.

Ideas

Ideas about society come from everywhere: everyday life, a novel, an unusual event, an analogy, a misunderstanding, a slip of the tongue, a silly joke. Ideas seem to appear more or less spontaneously. Most ideas turn out to be wrong or to be dead ends. For example, social scientists once thought that temperate climates caused higher forms of civilization to develop. As it turns out, this idea of climactic determinism does not do a very good job of explaining civilization. More than anything, this thinking showed that those living in temperate climates were ignorant of non-Western cultures and of the complexity of most cultural forms.

Good ideas, or at least those that stand up under scrutiny, become part of the stock of knowledge that is passed from one generation of scholars to the next. In social science, abstract knowledge about social life is called social theory. As it turns out, most people know a lot of social theory without studying it. They know, for example, that bureaucracies can become cumbersome and even choke on their own paperwork and procedures. They don't need to study organizational theory—a branch of social theory—to know this. They also know that most people most of the time act in ways to maximize their material gains and other self-interests. They don't need a theory of rational choice—another branch of social theory—to understand this. Still, social theory is valuable because this body of thinking explores these and other ideas in depth. What kinds of factors prevent bureaucracies from choking on their own procedures? Under what general conditions do people not make what seem to be obvious rational choices? Or, even more fundamental, is it always possible to tell which choices are rational and which are not?

The task of making sense of social life is daunting. The accumulated knowledge of social life represented in social theory offers an important resource. Some social research, as noted in Chapter 1, seeks to improve this body of knowledge by testing ideas derived directly from theory or by identifying general patterns that elaborate theoretical ideas. Not all research, however, is theory centered in this way. Social researchers who seek to interpret culturally or historically significant events, for example, view social theory as a reservoir of possible interpretations. Likewise, researchers who seek to give voice, another key goal of social research, recognize that their research cannot proceed without some theoretical guidance, yet their primary theoretical objective is to contribute to theory by learning more about phenomena and groups that have been ignored or misrepresented. However, even research that is more concerned with contributing new knowledge to this pool of ideas than with using existing ideas participates fully in the dialogue with ideas.

Analytic Frames

When most researchers approach the pool of ideas known as social theory, they usually have a specific research question or problem in hand. For example, a researcher might be interested in understanding why it is that people vote the way they do. What theoretical ideas (that is, ideas from the pool known as social theory) might help? Different ideas lead to different ways of framing and using evidence.

For example, one very simple theoretical idea is the notion that people act in ways that maximize their self-interests—they make rational choices. This theoretical idea sees the question of voting as an individual-level decision based on a sober assessment of a person's costs and benefits. The researcher would thus see the act of voting as a calculation of individual gains and losses given different outcomes, a calculation that would vary across individuals depending on their characteristics (for example, their income, their family size, and so on). In short, the idea of rational choice would lead the researcher to construct a particular analytic frame for understanding voting, which, in turn, would cause the researcher to see voting in a specific way. A different idea implemented through a different analytic frame might lead to a dramatically different view of voting, a different way of breaking it into its key components. For example, a theory that emphasized processes of social influence would turn the investigator's focus to the nature of each voter's social networks.

Thus, analytic frames are fundamental to social research because they constitute ways of seeing. While this notion may seem abstract, consider the operation of analytic frames in everyday life: As people go through their lives, they classify and characterize the things around them. For example, they know how to distinguish between "people standing around in a room" and "a party" because they understand and can use the term *party*. They also generally know what makes a party fun—which ingredients in what quantities, and so on—which is another way of saying they know how to characterize parties in different ways. Another way to describe people's understanding of parties is to say that they have an implicit *analytic frame* for parties. An analytic frame defines a category of phenomena (for example, parties) and provides conceptual tools for differentiating phenomena within the category (what makes them more and less successful, more and less formal, more this, less that, and so on). In short, analytic frames articulate ideas, in this case the *idea* of a party.

The person who is ignorant of the term *party* may not be able to tell the difference between a conference and a party. Both involve rooms full of people talking, often at the same time, often without listening to each other, often with laughter, and so on.

Now consider a related example from social research (Smith-Lahrman 1992), which further illustrates the frame as a way of seeing. In some coffeehouses people spend a lot of time *avoiding* interaction. They use posture and props such as newspapers and books to maintain social boundaries and social distance. In this sense their noninteraction is intentional and therefore is a social accomplishment. A quiet coffeehouse is not a social vacuum; it is teeming with purposeful social behavior.

Armed with the proper analytic frame—one emphasizing nonverbal communication—it is possible for social researchers to *see* that the noninteraction is accomplished. Without this frame, it might appear simply that "nothing is happening" when in fact massive efforts to achieve noninteraction are being exerted throughout the coffeehouse. In short, without a frame for accomplished noninteraction, researchers might be blind to its occurrences. They might also fail to consider similarities and differences among its occurrences across broad social spaces (for example, differences in how it is accomplished in trains, airports, elevators, and so on; differences in how men and women accomplish it; and other important considerations).

The process of using analytic frames to classify and characterize phenomena is carried out explicitly and formally in social research. Sometimes a social researcher will study something because it is unclear what it is or how it should be characterized. Is the movement toward "political correctness" a fad? Is it a social movement? Is it a new religion? Is the wave of anorexia among young women a response to fashion? Is it internalized misogyny? Is it an effort to erase gender differences by starving off secondary sex characteristics? Is it a form of mass inchoate protest against traditional gender roles—a hunger strike? Which analytic frames work best? A researcher may try several frames to see which makes the most sense of the phenomenon and leads to new insights.

Consider a more detailed example: Being a private in the army could be understood as a "job," and a researcher might study the work of a private as one might study the work of a unionized factory worker. Alternatively, a researcher might use the frame of "initiate." When people first join institutions that penetrate many aspects of their lives (for example, a religious commune, the priesthood, boarding schools, colleges and universities, and so on), they give up much of their old identities and adopt new ones. To study privates as "initiates" is to examine this process of identity change. The researcher who uses the analytic frame of "job" constructs a very different representation of privates than the one constructed by the researcher who uses the frame of "initiate." In fact, ambiguities about what it means to be in the military led two researchers (Moskos and Wood 1988) to publish the book *The Military: More Than Just a Job?*

By debating, using, and formalizing analytic frames, researchers are able to relate their research to the work of other researchers and to accumulate general knowledge about social life from their separate, individual efforts. For example, the researcher who uses the frame of "initiate" to study privates contributes to the body of knowledge concerned with basic mechanisms of identity change. The researcher who uses the frame of "job" contributes to the body of knowledge that addresses the organization and control of work.

Because analytic frames both classify and characterize social phenomena, they have two main components. When researchers use concepts to *classify* the phenomena they study, they **frame by case**. When they use concepts to *characterize* these cases, they **frame by aspect**. Both components of analytic frames are important parts of the dialogue of idea and evidence in social research.

Framing by case. When a social researcher states that most of what occurs in coffeehouses is "accomplished noninteraction," he or she classifies the phenomenon. In essence, the social researcher answers the question: What is this—the phenomenon being studied—a case of? The social life of a coffeehouse provides a case of accomplished noninteraction. Framing by case (that is, answering the question: What is this phenomenon a case of?) is an essential part of the process of social research (Ragin and Becker 1992).

When researchers claim that the people and events they are studying are an instance or "case" of something wider and more important, a larger category, they offer a frame for their research. For example, to argue that it is important to study the former Yugoslavia under Tito (the man who ruled Yugoslavia over most of its post–World War II history) as "a case of successful containment of ethnic conflict" is to frame this study as an instance of a more general category. Implicit in this statement is the idea that there are many such instances of "successful containment of ethnic conflict" and that the study of Yugoslavia under Tito should make a contribution to that general body of knowledge. Defining the case in conceptual terms—as an instance of something broader—is the most important part of the framing of a study. When more than one case is studied, they are often seen as multiple instances of the same larger category. For example, a comparative study of several instances of the successful containment of ethnic conflict might examine *specific* periods of the history of Yugoslavia, Malaya, Sri Lanka, Lebanon, the Soviet Union, India (when it was a colony of Great Britain), Canada, Israel, Belgium, Switzerland, and other countries (see Schermerhorn 1978; Rothschild 1981).

The broad conceptual categories that frame social scientific studies do not always involve large units like countries or abstract units like social interaction. The units can be almost any size. For example, a researcher might frame a study of the conflict between the pro-choice and pro-life movements as an instance of "polarized social movements." Another case of polarized social movements in the United States might be the conflict between organizations representing unions and those representing corporations in the conflict over "right-to-work" legislation.

Still smaller units are involved when a researcher frames fraternities and sororities as instances of "same-sex communal groups." And still smaller units are involved when interaction rituals like greetings are studied as instances of "efforts to cultivate relationships." All these examples involve framing by case. Even large-scale survey research involves framing by case. When a survey is used to examine the relation between economic interests and voting preferences, for example, the frame treats survey respondents as rational actors.

Framing by aspect. Specifying the broader category that is relevant to an investigation is only part of the process of analytic framing. Framing also involves specifying the key features or aspects that differentiate the cases in a broad category. Framing by case establishes an important category or set of phenomena; framing by aspect indicates how the cases within a category vary.

For example, social situations that qualify as sites of accomplished noninteraction (a category that includes coffeehouses, airports, buses, elevators, waiting rooms, some types of bars, and so on) vary in important ways. How do people accomplish noninteraction in all these different settings? What verbal, nonverbal, and other behavioral cues are used? What features of settings influence which cues are used and how they are used? The list of relevant aspects of settings that should be considered in this frame is very long. Sometimes noninteraction is accomplished among strangers, sometimes among acquaintances. The settings where it is accomplished vary by social density: sometimes people are spread out and can move about (as in an airport); sometimes they are tightly packed (as in a plane). Some social spaces are closed (buses, for instance); some are open (parks). Social settings that manifest high levels of accomplished noninteraction vary in many other ways, as well. Each of these features may have an important impact on how noninteraction is accomplished in each setting. Once social researchers answer the question "What is this a case of?" (that is, once they frame by case), they use theory and other ideas to identify the major features of cases in the frame and thus frame by aspect.

Consider again the study of the former Yugoslavia under Tito. To state that this is an instance of "successful containment of ethnic conflict" only partially frames this case. It is also necessary to elaborate the important aspects of the instances within this category. There may be many different ways of successfully containing ethnic tensions in nation-states, and each way of containing them may involve putting together a different combination of government policies and political strategies. Further, strategies that work well in some contexts may not work at all in other contexts. For example, it may be possible to appease an ethnic minority with a modest redistribution of resources if the ethnic minority is much poorer than the dominant ethnic group. This strategy might fail altogether if groups have roughly equal resources or if the minority is richer than the majority. Another example: If there are several minority groups, it may be possible to play them off against each other—to divide and conquer. Obviously, this strategy cannot be pursued if there is only a single minority group or one large minority group and several smaller minority groups. In short, there are many different aspects to the "successful containment of ethnic conflict." The researcher's analytic frame for the study of the containment of ethnic tensions should embrace all these aspects.

Framing by aspect helps social researchers see both what is present and what is absent in a given case. For example, assume that the analytic frame for "successful containment of ethnic conflict" is applied to Yugoslavia under Tito. This frame both guides the researcher to examine specific phenomena that were present in Yugoslavia (for example, the government's effort to buy off ethnic conflict through a redistribution of national resources), but also to consider the impact of features that were *absent* in this case (for example, multiparty democracy) but *present* in other cases covered by the analytic frame (Belgium, for instance). Would a redistribution of resources have dampened ethnic antagonism if Yugoslavia had been a multiparty democracy during this period? Would redistribution have been feasible under these conditions?

In all social research some guide is needed to see what is present and what is absent in a given case. Sometimes, the things that are absent in a case help the most in explaining why it is one way and not another. Note however, that it is easy to miss what is absent without some sort of analytic frame to guide the analysis. Without this guidance, the tendency is to focus only on what is present.

Together, framing by case and framing by aspect constitute two key conversations that take place in the dialogue of ideas and evidence. How and when these conversations take place differ greatly from one research strategy to the next (Diesing 1971). Sometimes the analytic frame for a research project exists before the research begins and structures most aspects of the research; sometimes the frame is articulated in the course of the research. The interplay of analytic frames and research strategies is addressed in the final section of this chapter.

Evidence

When most people think about social scientific evidence, they usually think of questionnaires and telephone surveys. After all, social scientists conduct huge surveys on all aspects of social life and then publish their findings—the percentage of people who think this or that or who do this or that, broken down by gender, race, age, education, income, or whatever. However, social scientists are not limited to survey data. In fact, only a relatively small proportion of social scientists are survey researchers. Many study phenomena that cannot be addressed with questionnaires.

All facets and features of social life offer evidence; virtually everything to a social scientist is "data," at least potentially. Some social researchers observe social life as it occurs in everyday settings. They take reams of field notes on people's daily routines of family, work, and play in their various locales: street corners and kitchens; offices and factories; country clubs, churches, bars, back alleys, and emergency rooms. Others conduct in-depth interviews with people from different walks of life and try to stimulate their subjects to be more introspective about their lives, to analyze their own thoughts and actions. A researcher interested in labor control, for example, might interview fifty employees of a factory, drawn from all levels and divisions of its work force. Other researchers study past events, using historical documents and records from libraries and archives. Still other researchers study patterns across whole cities and countries, using official statistics published in the reports of government and international agencies. There are many, many sources of evidence about social life, and social researchers have explored virtually every type.

Not only are there many different sources of data, each instance of social life potentially offers an infinite amount of information. The **empirical** world is limitless in its detail and complexity. Social research thus necessarily involves a **selection** of facts. Most facts must be ignored as irrelevant; otherwise, research would be impossible.

Consider the seemingly simple task of taking notes on what occurs in a classroom during an hour-long lecture. First of all, it's necessary to set the stage properly—a physical description of the lecture hall, its atmosphere, the number of people in attendance, their distribution in the lecture hall, and so on. This description could easily fill one notebook. Next, there is the lecture itself. Exhaustive notes on the content of an hour-long lecture could fill another notebook. But then there's also the lecture as a performance, which includes nonverbal behavior (for example, gestures and other bodily movements) and the interplay of the verbal material and nonverbal behavior. This information could easily fill several notebooks. There should also be notes on the reactions of students in the audience. Of course, with enough resources it would be possible to monitor the behavior of each person throughout the hour, including their verbal and nonverbal behavior, their note taking, their social interaction, and so on. This would yield enough information to fill at least one notebook for everyone in attendance. And don't forget that it's also possible to take notes on the interaction between the lecturer and the cues—verbal and nonverbal, conscious and unconscious—that the listeners send to the lecturer. A videotape of this interaction could be studied for many years and yield many more reams of field notes. In short, to try to capture the full details of social life, even a very small slice of it, is a colossal undertaking.

Because every slice of social life potentially offers an unlimited amount of evidence, researchers must be selective in their use of evidence. It would take an infinitely long research report to use all the evidence a typical case offers. Although social researchers usually collect large volumes of evidence, the quantity they collect can, at best, constitute only a tiny fraction of the evidence they *potentially* could collect. They try to focus on only the most significant portions, using their ideas, analytic frames, interests, past studies, and so on to help them assess what seems most important to their research questions. The problem of selecting evidence returns us to ideas and analytic frames. Without some sort of sensitizing ideas or concepts, the world seems an amorphous blob. We perceive evidence and select some of it as especially relevant because of our ideas and frames. As will become evident in the next section, however, the images social scientists construct from these bits of evidence may not conform to the initial ideas and frames that defined the evidence as relevant in the first place.

This need for selectivity introduces a problem. When a writer becomes an advocate for a particular point of view, he or she "selects" for reporting only the bits of evidence that support that position. This kind of selectivity involves an ignorance, either willful or unconscious, of evidence that favors opposing points of view. Ignoring evidence is not always willful, however; sometimes it is a product of limited awareness or limited resources and thus is unintentional. For example, before the rise of feminist perspectives in the social sciences, many researchers did not see the pervasiveness of sexism in everyday life. Thus, evidence bearing on sexism was often missed in studies of a wide range of social relations. Many other forms of ignorance and unrecognized bias infect all research. While it would be great if every social scientist had some way to recognize the impact of such bias on his or her own research, there is no automatic safeguard. Social scientists are only human, and they can't designate evidence as relevant if their unrecognized biases persuade them to ignore it.

The only real safeguard to unrecognized bias is the fact that social science is communitarian (Merton 1973). Social scientists write for other social scientists and they judge each other's work. They try to detect bias. Almost every social scientific representation of social life is evaluated by other social scientists before it is published or made public in some way, and it is usually subjected to close scrutiny after it is published, as well. In fact, social scientific representations are subjected to more scrutiny than most other representations of social life. Of course, if all or even most social scientists share the same unrecognized biases, as is sometimes the case, then the influences of biased selection of evidence will not be immediately recognized. However, social scientists believe that future generations of social scientists will uncover and correct the unrecognized biases of preceding generations.

Images

Ideas and analytic frames direct the researcher's attention to specific kinds and categories of evidence. From an ocean of potential data, the researcher selects what seem to be the most relevant portions. Once a sufficient body of relevant evidence has been collected, the researcher's next task is to make sense of it and at the same time relate it back to the ideas and frames that initially motivated the collection of evidence.

Researchers make sense of their evidence by constructing images of their cases from the data they have collected. In effect, an image is constructed by the investigator when he or she brings together or synthesizes evidence. Images often imply motives or say something about causation. When a researcher notes that people with more income tend to vote for the Republican Party, for example, he or she creates part of an image of how a preference for Republicans comes about. Thus, an image is the product of the effort to bring coherence to data by linking bits of evidence in meaningful ways.

Consider an extended example: The researcher who wants to understand how medical students become doctors may start the research with specific ideas about the professions and the nature of professional socialization. One common notion is that each profession upholds certain values or principles and that professional socialization involves learning how to apply these principles in everyday situations. For the medical profession one central value might be that the health of the patient comes before all else. Because this analytic frame emphasizes the application of abstract principles, the researcher might initiate data collection by observing medical students in clinical practice, with special attention to whatever general principles seem important in these settings. A few weeks of fieldwork in the clinics of a teaching hospital would no doubt result in a huge volume of notes on what was observed. What images of medical students and their professional training emerge from this fieldwork? Which images make the most sense of this new body of evidence? Which aspects of the professional socialization of medical students should be investigated next?

Images are formed from evidence in order to make sense of the evidence, summarize it, and relate it back to the ideas that first motivated the collection of evidence. To construct images researchers connect different aspects of cases to form coherent portraits. Suppose the researcher studying medical students found that clinical decision making revolved not necessarily around the best interests of patients, but mostly around the needs of doctors and hospital officials to protect themselves from charges of malpractice. The *image* of professional socialization that emerges from this connection is that training centers on getting medical students to exaggerate the correspondence between this need for protection from malpractice charges, on the one hand, and the "best interests" of patients, on the other. After all, charges of malpractice can be avoided in part by exercising extraordinary caution—for example, by ordering many laboratory tests on each patient so that every possible diagnosis is covered. This excessive use of laboratory tests could be construed as "thoroughness" or "expert care" and thus "in the patient's best interests," even though testing is often invasive, unpleasant, expensive, and may cause serious reactions and even secondary illnesses.

This image of professional socialization, built up from evidence, both elaborates and challenges the initial frame. The initial frame emphasized the importance of abstract professional values in professional socialization (for the medical profession: "putting the patient first"). The image constructed from evidence, however, indicates that, in everyday settings, professional values are learned primarily in the context of practical and institutional concerns (for example, avoiding charges of malpractice). In other words, practical and institutional concerns *modify* how professional values are understood and implemented. This image of the training of medical students, built up from observations of decision making in clinics, organizes the evidence the researcher has collected in a way that highlights its relevance to the original analytic frame.

Consider another example of images in social research. Researchers have noted that many inner-city neighborhoods have lost their middle class families to more prosperous, outlying neighborhoods and that these losses have accelerated the decline of inner-city neighborhoods (Wilson 1980, 1987). This connection between the loss of middle class residents and accelerated neighborhood decline contrasts two images. The first is a "thriving minority community"—what it presumably was like before the flight of the middle class: a neighborhood composed of individuals with different income levels (poor, working class, and middle class), with the more successful members offering community leadership, role models, information on how to get ahead, jobs in locally owned businesses, and many other resources for less fortunate members. The second image—the post-flight community—is an "inner-city ghetto" and offers a striking contrast: uniformly poor members with high rates of unemployment, crime, violence, drug addiction, welfare dependence, teenage pregnancy, female-headed households, despair, and so on. Linking these two images is the "exodus" of the minority middle class. This example of the construction of images can be used to illustrate three of their important qualities:

1. Images are **idealizations** of real cases. Every real neighborhood is complex and ever-changing. It is doubtful that any neighborhood perfectly fits either of the two images just elaborated, the "thriving minority community" or the "inner-city ghetto," at least not for any great length of time. Images are exaggerations because they are necessarily constructed from selected pieces of information; they cannot reproduce real cases because real cases are infinitely detailed and complex. Thus, images should be seen as pure or idealized cases (Weber 1949). These two terms—*idealized* (as in idealized cases) and *idealization* (the process)—are used here not to indicate desirability, as in the statement: This area offers an *ideal* climate for year-round outdoor sports. Rather, they are used to indicate that images are *abstractions*. Unlike theoretical ideas, however, they are abstractions that have a specific grounding in a body of evidence. The process of constructing idealized cases (idealization) involves abstracting from

information about empirical cases to conceptually elaborated images. As idealized cases, images can be linked to theoretical ideas expressed in analytic frames.

2. Most images imply or embody *explanations*. Most explanations are *causal*, which means simply that they offer accounts of why things are the way they are, emphasizing connections among different phenomena. When we explain the accelerating decline of inner-city neighborhoods by pointing to the exodus of the minority middle class, we pinpoint a causal connection. The key part of a causal explanation is its *cause*-words. Cause-words, like "exodus," are the most important part—the action part—of the images that social scientists construct. *Exodus* connotes collective, willful abandonment of a specific locale. It's packed with meaning. Words like *exodus* link images to analytic frames, ideas, and ultimately to social theory. There are social scientific theories, for example, that seek to conceptualize the variety of push-and-pull factors that cause people to move from one community to another. These theories are relevant to many kinds of migrations: the exodus of minority middle class people from inner-city neighborhoods, the gentrification of other urban neighborhoods, and the back-and-forth migration of Mexicans to and from particular villages in Mexico and specific communities in the United States.
3. Images are *guides* to further research; they suggest new research questions and new avenues to explore. Images help researchers see what they might otherwise miss and thus lead them to examine social life in a more systematic way. For example, we can ask: Have we omitted any important aspects in either of these two images? For instance, do most "inner-city ghettos" also lack grass-roots political organizations? Are there important differences between those with such organizations and those lacking such organizations? Another example: Are there inner-city minority neighborhoods with a good cross section of income groups (poor, working, and middle class) that nevertheless developed high rates of crime, violence, drug addiction, teenage pregnancy, and so on? If so, why didn't the existence of middle class role models, leaders, and so on forestall these developments? Still another example: Would the return of middle class minority members to an inner-city ghetto help roll back the rising tide of violence, drug addiction, welfare dependence, and so on? These questions follow directly from the two images constructed.

Once formed, images interact with analytic frames. The process of constructing images (or **imaging**) complements the process of deriving analytic frames from theory (or framing by case and framing by aspect). While these two activities, framing and imaging, seem to correspond to deduction and induction, it would be a mistake to limit them in this way. Even though imaging is mostly inductive, it uses evidence that has been defined as relevant by the ideas and frames the researcher brings to the study. It is difficult to form an image from evidence without first using some sort of initial analytic frame to highlight or define relevant evidence.

Likewise, even though framing is mostly deductive, the body of knowledge from which frames are derived summarizes accumulated evidence-based knowledge about social life. Thus, framing is based on a vast body of systematized evidence. Furthermore, at the start of most research projects, the analytic frame for the research is usually only half developed, at best. Social theory is abstract, general, and often vague, so much so that several different frames can be derived from the same set of ideas. In the course of the research, if the images formed from evidence are compatible with the initial analytic frame, then they can be used to clarify and refine it. However, sometimes the images formed from evidence reject the initial framing and force the investigator to seek out or develop new frames (Walton 1991, 1992). This interaction between images and frames is best understood as a process of *retroduction*, a termed developed by philosophers of science (Hanson 1958) to describe how induction and deduction work together in research (see Figure 3.1 and Chapter Four).

Representations

The dialogue of ideas and evidence culminates in representations of social life (see Figure 3.1). In social research, analytic frames and images interact to produce a progressively refined portrait or picture, which becomes the representation (and the explanation) that the researcher offers.

In many ways social scientific representations can be compared to photographs. The photographer selects an image to be represented, taking care to ensure that the right elements are brought together in the image. By bringing together these elements, the photographer conveys the message or ideas he or she intends. The image in the photograph is framed in several ways. Within the photographic image itself, it is framed by focus—some parts of the image are foregrounded and the focus is sharp, others are backgrounded and out of focus. The photographic image is framed as well by its boundaries. It can be cropped in a variety of ways; each cropping has a different effect on the meaning of the image. Consider the fact that the world around the photographic image is seamless—it goes on forever. The frame established by the photographer limits the context of the image. Images are unclear if they are not properly framed.

So it is in social research. The main part of the representation is the image, which is built up from evidence. Researchers link pieces of evidence together to make images. The analytic frame provides the context for creating and understanding the image, establishing conceptual boundaries around the evidence-based image. It is important to understand that in both social research and photography, representations appear to audiences as finished products, complete with images and frames. However, these finished products result from a long process. There is an interplay of possible frames and potential images in the construction of every representation.

At the core of every social scientific representation is an explicit or implicit explanation of some major aspect of the phenomenon it represents. The explanation is what gives the representation coherence because it is very difficult to "tell about" social life (that is, represent it in some way) without giving some kind of account of it (that is, explain it). For example, the researcher who studies interaction in a coffeehouse explains how people accomplish noninteraction; the researcher who studies ethnic tensions in a range of countries explains how conflict may be prevented or at least postponed; the researcher who studies medical students explains how they come to see a correspondence between the practical concerns of doctors and hospitals and their professional commitments to patients; finally, the researcher who studies inner-city neighborhoods explains how their loss of middle class members contributed to their decline.

Ways of representing the final product of the interaction of frames and images in social research are varied, and the intended audience for a representation has a strong impact on how it is presented. While it is possible to imagine a variety of ways of representing the results of social research (for example, documentary films, dramatic performances, text mixed with still photographs and sound recordings, multimedia presentations, and so on), social researchers tend to use academic books, journal articles, textbooks, and an occasional mass circulation article. In other words, they use traditional text media almost exclusively. Within each of these media, however, different formats may be employed: tables, charts, equations, transcripts, narratives, vignettes describing typical or exemplary cases, and so on.

Processes and Strategies of Social Research

While all social research involves interaction between images and analytic frames, the nature of this interaction can differ significantly from one research project to the next. A key consideration in understanding these differences is the role of analytic frames in research. In some research frames are **fixed** at the start of the study, while in others they may be either **flexible** or **fluid** and change during the course of the investigation.

Analytic frames may be elaborated at the outset of a research project and remain more or less the same throughout the study. This use of fixed analytic frames is often necessary, for example, in studies that seek to test theories. In essence, the analytic frame implements a hypothesis to be tested. If images constructed from the evidence are inconsistent with the hypothesis, then the hypothesis is rejected. Fixed frames are also common in research that seeks to make predictions based on current trends and in studies that seek to document broad patterns.

Fixed frames are most compatible with quantitative research on covariation (see Chapter 6). In research of this type, there is sometimes a close correspondence between the analytic frame developed at the outset of the research and the data set that the researcher then constructs. Recall that analytic frames elaborate ideas by specifying both a category of phenomena and the major ways phenomena within the category vary. For example, a frame that looks at voters as rational actors sees voters as the category and their individual-level differences (for example, their different educational levels, income levels, and so on) as aspects that might explain their different choices. This analytic frame readily translates to a survey format, where potential voters are queried about their demographic characteristics and their voting behavior (see Page and Shapiro 1991). This simple translation from the analytic frame to survey data permits a direct test of the idea that first inspired the frame—that voters make rational choices. If the images constructed from the data do not correspond to the idea of rational choice, then the hypothesis is rejected.

In other studies, the analytic frame is flexible; it is elaborated as a guide for research, showing which kinds of factors might be relevant in which contexts. A flexible frame is useful, for example, in studies that seek to explore diversity or advance theory. A flexible frame shows the researcher where to look and what kinds of factors to look for without making specific hypotheses about relationships among factors.

Flexible frames are common in comparative research (see Chapter 5), especially when the goal is to make sense of a range of diverse cases. Consider a researcher who is interested in tyranny and explores it by studying many of the major tyrants of the twentieth century (for example, Stalin, Hitler, Duvalier, Trujillo, Sadam Hussein, and so on; see Chirot 1994). The analytic frame might direct the researcher to examine a range of factors: how these tyrants came to power; what good, if any, they accomplished; who supported them, both domestically and internationally; what ideologies they used, if any, to justify their cruelty; how much suffering they caused; and so on.

Examination of this evidence might lead the researcher to differentiate types of tyrants. For example, the evidence might show that the more ideological tyrants (for example, Hitler and Stalin, among others) caused more suffering, while the less ideological tyrants caused less suffering. In this way, the researcher could elaborate the analytic frame, used initially as a way to guide the research, with these evidence-based images (the two main types of tyrants—more ideological and more abusive versus less ideological and less abusive). In this way, the research could offer important leads for the advancement of theories of political oppression (Chirot 1994).

Finally, in some research, analytic frames are fluid. Researchers who seek to give voice (one of the goals of research discussed in Chapter 2), for example, may want to limit the influence of preexisting ideas. Of course, they must have some initial ideas about their research subjects; otherwise, the research could not be started. But these ideas might be quickly set aside once the research is underway. Alternatively, the researcher might start with several frames, and move fluidly among them depending on the nature of the evidence as it accumulates. The use of multiple, fluid framing is especially appropriate when researchers seek to give voice because a fixed analytic frame might prevent researchers from hearing the voices of the people they study.

Fluid frames are most common in qualitative research (see Chapter 4). Often researchers will not know that their case is "a case of" when they first start their investigation. When there are many possible framings, each can be explored—to see which help make the most sense of the evidence. Sometimes, multiple frames are retained throughout a project and included in the representation, especially if these different framings illuminate the subject in complementary ways. The American Civil War can be framed in many different ways, as a fight over slavery, as a fight over states' rights in a federal system of government, as a struggle between a plantation society and an emerging industrial society, and so on. These different frames can be integrated into a single, encompassing portrait.

Framing a case in different ways enriches the understanding of the case when each framing offers insights for other framings. When different framings contribute insights for other framings, the case or cases that are the focus of the study are said to be "rich" because they provide so much raw material for the advancement of social thought. Unfortunately, this creative interaction among frames is relatively rare in social research. Typically, in qualitative research investigators struggle simply to come to terms with their cases. Existing frames may not work well at all, and the case becomes a platform for developing new ideas and new frames.

The Challenge of Social Research

Ideas and evidence are everywhere. It's no great surprise then that there are so many people busy constructing representations of social life, from poets and painters to playwrights and political scientists. Different ways of constructing representations require different kinds of regimen. The regimen of poetry, for example, is to construct representations that make the most of as few words as possible. The regimen of social research is also strict, though quite different, and it is reinforced by the primary audience for social research—social scientists.

The regimen of social research demands both clear specification of the ideas that guide research and systematic examination of the evidence used to build images and representations. The challenge of social research is to construct powerful and instructive representations of social life that contribute to the ongoing conversations about social life called social theory and at the same time embrace a breadth or depth of evidence about social life in a systematic way. This challenge can be met by building a dialogue of ideas and evidence—analytic frames and evidence-based images—into the process of social research.

PART II

Strategies of Social Research

The diverse goals of social research favor strikingly different research strategies. The goal of identifying general patterns, for example, suggests a research strategy appropriate for the study of many cases. However, when many cases are studied, it is difficult to study each case in depth. Typically, only a relatively small number of features across many cases can be incorporated into a representation. Thus, studies that examine many cases tend to focus on select aspects of cases (that is, on a specific set of "variables"). By contrast, consider the goal of assessing historical or cultural significance. Usually, this strategy focuses on a small number of cases, sometimes only one, and researchers examine each case in great detail. This intensive strategy attempts to construct a full portrait through close analysis of the links among many different aspects of cases. Studies that focus on a small number of cases tend to examine many features of those cases.

More generally, there is a trade-off between the *number of cases* and the *number of features of cases* social researchers typically can study and then represent. At one extreme is most qualitative research: few cases, many features. At the other extreme is most quantitative research: many cases, few features. In between these two extremes is comparative research. The comparative study of diversity across a moderate range of cases strikes a balance between in-depth knowledge of cases and broad knowledge of relations among variables (Ragin 1991). It is the best strategy when there are too many cases for close, detailed investigation of each case, but too few for quantitative analysis.

These three strategies are the primary concern of Part II of *Constructing Social Research.* Chapter 4 presents the qualitative study of commonalities as a strategy that is best suited for clarifying categories and concepts. Chapter 5 presents comparative research as a strategy oriented toward identifying and unraveling complex patterns of similarities and differences. Chapter 6 discusses quantitative research as an approach that most often focuses on the covariation of features across many cases.

Each of these strategies offers a different approach to evidence, and each has core data procedures that investigators use when they construct images from evidence. These different data procedures structure the interaction between researchers and their evidence, allowing them to digest large quantities, pinpoint decisive bits, or identify subtle patterns. Procedures differ from one strategy to the next because the nature of the evidence and the goals of research can vary so greatly. After all, a research strategy that uses a relatively small amount of information on each of 1,000 cases calls for data procedures that differ dramatically from one that involves sifting through mounds of detailed information on a small number of cases.

Of course, all researchers work with evidence in many different ways—they use whatever procedures help them make sense of the evidence they collect. However, it is easier to grasp the three strategies discussed in Part II if each is presented along with the data procedures that are most compatible with the goals and logic of that strategy.

Qualitative methods are appropriate for in-depth examination of cases because they aid the identification of key features of cases. Most qualitative methods *enhance* data. Comparative methods are appropriate for the study of diversity because they are sensitive to complex differences among cases. Comparative methods can be used to *elucidate* subtle patterns in the data. Finally, quantitative methods are appropriate for the study of relationships among variables because these methods can be used to assess the correlation between two or more features across many cases. Quantitative methods can *condense* data on the relationship between two variables across many cases into a single number.

While the differences between these three strategies are striking, it is important to remember that they all involve a dialogue of ideas and evidence, as described in Chapter 3. Further, all three approaches are used to construct social scientific representations of social life, the primary goal of social research.

4

Using Qualitative Methods to Study Commonalities

Introduction

In some respects, qualitative research does not seem as scientific as other kinds of social research. Usually when we think of social science we think of sweeping statements like: "People with more education tend to get better jobs." "Poor countries tend to have more social conflict and political instability than rich countries." These statements offer "big-picture" views that say nothing about individual cases. In these big-picture views, a single statistic or percentage can summarize a vast amount of information about countless cases.

But a lot may be missed in the big picture. Often, researchers do not want these broad views of social phenomena because they believe that a proper understanding can be achieved only through *in-depth* examination of specific cases. Indeed, qualitative researchers often initiate research with a conviction that big-picture representations seriously misrepresent or fail to represent important social phenomena.

Consider the researcher who wants to understand the fascination that some people have with guns—for example, gun collectors, some military personnel, hunters, and other enthusiasts. A big-picture view might show that certain categories of people (for example, lower middle-class white males) are more likely to collect guns and subscribe to magazines devoted to guns (Stinchcombe et al. 1980 study this question). But does the big-picture view really say very much about the fascination with guns? What's the best way to study and understand this fascination?

A lot can be learned simply by talking to gun enthusiasts. They can be located in gun shops, gun clubs, and at practice ranges. The researcher in this case might try to get to know as many as feasible and interview them in depth. How did they get started with guns? How many guns do they own? How often do they shoot them? How do they feel when they are using them? How do they feel when they don't have easy access to a gun? How many of their friends are gun enthusiasts? Do they feel that law enforcement agencies are effective? What do they think about capital punishment? What political organizations, if any, do they belong to? How do they vote?

From these interviews it would be possible to build an image of at least one major type of gun enthusiast, to craft a composite image based on interviews of many individuals. This composite image could be fleshed out further by studying the magazines and other literature that the interviewees read and by observing what goes on at social gatherings of gun enthusiasts. The key would be to achieve as much in-depth knowledge as possible and look for common patterns among gun enthusiasts and their social worlds.

Sometimes the emphasis of the qualitative approach on in-depth knowledge means the researcher examines only a single case (for example, the life history of a single individual or the history of a single organization). Knowing as much as possible about one case is not easy because every case potentially offers information that is infinite in its detail. Much of this information is not useful because it is redundant or irrelevant, given the researcher's questions and purposes. In the qualitative approach, researchers must determine which information is useful in the course of the investigation, and they become more selective as additional knowledge about each case is gained. In the course of learning more about the research subject, the investigator sharpens his or her understanding of the case by refining and elaborating "images" of the research subject and relating these to analytic frames (see Chapter 3). These emerging images serve to structure further inquiry by marking some data collection paths as promising and others as dead ends.

Qualitative research often involves a process of *reciprocal clarification* of the researcher's image of the research subject, on the one hand, and the concepts that frame the investigation, on the other. Images are built up from cases, sometimes by looking for similarities among several examples of the phenomenon that seem to be in the same general category. These images, in turn, can be related to concepts. A concept is a general idea that may apply to many specific instances. Concepts offer abstract summaries of the properties shared by the members of a category of social phenomena. They are the key components of analytic frames, which, in turn are derived from ideas—current theoretical thinking about social life (see Chapter 3).

Consider a simple example first mentioned in Chapter 1: "Emotion work" is a concept developed by Arlie Hochschild (1983) to describe the conscious manipulation of feeling to create a publicly observable facial and bodily display. This concept, in effect, summarizes a lot of what flight attendants do because they often have to create certain appearances for passengers. Her study of flight attendants thus involved a mutual clarification of the *category* "the work of flight attendants" and the *concept* of "emotion-work." She refined the image of the flight attendant (an empirical category) as she clarified the concept of emotion work. This process of reciprocal clarification is ongoing and culminates in the representation of the research that the investigator offers at the conclusion of the study. The newly refined concepts—those that were elaborated in the course of the study—are featured in the representation of the results of qualitative research.

The Goals of Qualitative Research

Because of its emphases on in-depth knowledge and on the refinement and elaboration of images and concepts, qualitative research is especially appropriate for several of the central goals of social research. These include giving voice, interpreting historical or cultural significance, and advancing theory.

Giving Voice

There are many groups in society, called marginalized groups by social scientists, who are outside of society's mainstream, for example—the poor, sexual minorities, racial and ethnic minorities, immigrant groups, and so on. Often, these groups lack voice in society. Their views are rarely heard by mainstream audiences because they are rarely published or carried by the media. In fact, their lives are often misrepresented—if they are represented at all.

Techniques that help uncover subtle aspects and features of these groups can go a long way toward helping researchers construct better representations of their experiences. By emphasizing close, in-depth empirical study, the qualitative approach is well suited for the difficult task of representing groups that escape the grasp of other approaches.

Interpreting Historically or Culturally Significant Phenomena

How we think about an important event or historic episode affects how we understand ourselves as a society or as a nation. For example, in the middle to late 1800s the United States was involved in a series of territorial struggles with Mexico. These struggles can be interpreted as part of the inevitable westward expansion of European-Americans across a vast, sparsely populated continent. Or perhaps they can be seen as part of a pattern of unjust bullying of a generally peaceful neighbor. As the United States gains an ever larger Hispanic population, a revision of our understanding of these territorial struggles may help us adjust our view of the diverse collection of people who make up American society.

Methods that help us see things in new ways facilitate this goal of interpreting and reinterpreting significant historical events. Of course, if the evidence does not strongly support a new image, or offers better support to existing images, then new ways of understanding past events will not gain wide acceptance. The important point is that the qualitative approach mandates close attention to historical detail in the effort to construct new understandings of culturally or historically significant phenomena.

Advancing Theory

There are many ways to advance theory. New information about a broad pattern that holds across many cases (for example, a strong correlation; see Chapter 6) can stimulate new theoretical thinking. However, in-depth knowledge—the kind that comes from case studies—provides especially rich raw material for advancing theoretical ideas. When much is known about a case, it is easier to see how the different parts or aspects of a case fit together.

For instance, it is difficult to know how the structure of a nun's daily routines of prayer, work, and community life help her maintain her deep religious commitments without collecting detailed observations of the lives of nuns. This in-depth knowledge is useful for elaborating concepts such as "commitment" and for direct examination of the connections among the phenomena that the researcher believes illustrate and elaborate the concept, for example—the daily routines of those with strong commitments.

The value of qualitative research for advancing theory also follows directly from practical aspects of this type of research. It is impossible to decide which bits of evidence about a case are relevant without clarifying the concepts and ideas that frame the investigation. The initial goal of knowing as much as possible about a case eventually gives way to an attempt to identify the features of the case that seem most significant to the researcher and his or her questions. This shift requires an elaboration and refinement of the concepts that initially prompted the study or the development of new concepts. Researchers cannot forever remain open to all the information that their cases offer. If they do, they are quickly overwhelmed by a mass of indecipherable and sometimes contradictory evidence.

Finally, qualitative research also advances theory in its emphasis on the commonalities that exist across cases. In some studies cases may be selected that at first glance may seem very different. Identifying commonalities across diverse cases requires that the investigator look at the cases in a different way and perhaps discover new things about them. Diane Vaughan's study *Uncoupling* (1986), for example, focused not only on the breakup of conventional relationships—heterosexual marriages—but also on the breakup of homosexual relationships. Despite profound differences in the sexual orientations of her subjects, Vaughan found striking similarities in the process of "uncoupling" across these different kinds of relationships. By looking for similarities in unexpected places, social researchers develop new insights that advance theoretical thinking.

The Process of Qualitative Research

Qualitative research is often less structured than other kinds of social research. The investigator initiates a study with a certain degree of openness to the research subject and what may be learned from it. Qualitative researchers rarely test theories. Instead, they usually seek to use one or more cases or categories of cases to develop ideas. The qualitative researcher starts out by selecting relevant research sites and cases, then identifies "sensitizing concepts," clarifies major concepts and empirical categories in the course of the investigation, and may end the project by elaborating one or more analytic frames.

Selecting Sites and Cases

Qualitative research is strongly shaped by the choice of research subjects and sites. When the goal of the research is to give voice, a specific group is chosen for study. When the goal is to assess historical or cultural significance, a specific set of events or other slice of social life is selected. When the goal is to advance theory, a case may be chosen because it is unusual in some way and thus presents a special opportunity for the elaboration of new ideas.

Sometimes, however, cases are chosen not because they are special or unusual or significant in some way, but because they are typical or undistinguished. A researcher interested in medical schools in general, for example, might select a school that is typical or average, not the best medical school in the country or the worst (see Becker et al. 1961). To select a school at either extreme might limit the value of the study for drawing conclusions about medical schools in general. In short, because qualitative researchers often work with a small number of cases, they are sometimes very concerned to establish the *representativeness* of the cases they study (see Chapter 1).

In-depth knowledge is sometimes achieved through the study of a single case. Often, however, it is best achieved by studying several instances of the same thing because different aspects may be more visible in different cases. Consider a study of a neighborhood with many new immigrants. The researcher might find that in this neighborhood the cultivation of interpersonal networks (that is, making connections with lots of different people) is the key to the successful adjustment of immigrants to the United States. Much can be learned from studying one such neighborhood in depth. In fact, it is only through in-depth study that immigrants' use of interpersonal networks could be thoroughly documented. However, the study could be deepened further through the study of several immigrant neighborhoods. There may be various ways of establishing a reliable interpersonal network, depending on the cultural backgrounds of the immigrants. Different ways of establishing interpersonal networks might be more apparent in other neighborhoods.

When qualitative researchers collect data on many instances of the phenomenon under study, they focus on what the different instances have in common. Examining multiple instances of the same thing (for example, interviewing thirty-five flight attendants) makes it possible to deepen and enrich a representation (for example, a representation of the emotion work required in service jobs). A study of environmental activists might focus on the life experiences they share. A study of Catholic priests might focus on how they maintain their religious commitments. A study of immigrant neighborhoods might focus on the different ways of establishing and using interpersonal networks to facilitate immigrants' adjustment to their new surroundings.

When many instances of the same thing are studied, researchers may keep adding instances until the investigation reaches a point of **saturation**. The researcher stops learning new things about the case and recently collected evidence appears repetitious or redundant with previously collected evidence. It is impossible to tell beforehand how many instances the researcher will have to examine before the point of saturation is reached. In general, if the researcher learns as much as possible about the research subject, he or she will be a good judge of when this point has been reached.

Of course, if the cases selected for study are not sufficiently representative of the category the qualitative researcher hopes to address, then the point of saturation may be reached prematurely. A study that seeks to represent the work of taxi drivers in New York City may reach saturation (no new things are being learned) after the researcher interviews ten taxi drivers who are recent immigrants from Romania. However, these ten Romanian taxi drivers are probably not representative of all New York taxi drivers. The researcher should seek out taxi drivers with different backgrounds.

Even when qualitative researchers study many instances of the same thing (as when fifty priests are interviewed, for example), they often describe the case as singular ("the *case* of Catholic priests") because the focus is on commonalities—features that the instances share. By contrast, a quantitative researcher (see Chapter 6) interested in systematic differences (say, the covariation between age and strength of religious commitments among these same priests) would emphasize the fact that the research summarizes information on *many* cases (fifty priests). Statements about patterns of covariation (for example, "older priests appear to be more committed than younger priests") are more likely to be accepted if they are based on as many cases as possible.

This distinction is subtle but very important. The qualitative researcher who interviews fifty priests seeks to construct a full portrait of "the priest" and how priests maintain their deep religious commitments. It may be that the images that emerged changed very little, if at all, after the tenth priest was interviewed, and not much was learned from the remaining forty priests. The difference between ten and fifty is not important; what matters is the soundness of the portrayal of *this* case (the Catholic priest). If a study is done properly and is based on a sufficient number of interviews, it can be used for comparison with other cases (for example, comparing priests with the ministers of a Protestant denomination). The important point is that even though many examples of the same thing may be examined, research that emphasizes similarities seeks to construct a single, composite portrait of the case.

Use of Sensitizing Concepts

It is impossible to initiate a qualitative study without some sense of why the subject is worth studying and what concepts might be used to guide the investigation. These concepts are often drawn from half-formed, tentative analytic frames, which typically reflect current theoretical ideas. These initial, *sensitizing concepts* get the research started, but they do not straitjacket the research. The researcher expects that these initial concepts, at a minimum, will be altered significantly or even discarded in the course of the research.

A researcher studying hospital patients may bring "social class" as a sensitizing concept to the research and expect to find that patients from families with more income receive better care. However, the concept of social class, as expressed in family income, might prove to be too limiting as a frame for the research and be supplanted by an emphasis on some other aspect of family social status, such as occupational prestige of the head of the household. Sometimes concepts that seem important or useful early in the study prove to be dead ends, and they are discarded and replaced by new concepts drawn from different frames. Armed with these new concepts, the researcher may decide that some of the evidence that earlier seemed irrelevant needs to be reexamined.

For example, John Walton (1991, 1992) studied the conflict over water rights in Owens Valley, California, a struggle that pitted the residents of Owens Valley against water-hungry Los Angeles. (This struggle provided the background for the movie *Chinatown*, starring Jack Nicholson.) The battle over water rights dragged on for decades and generated so much mass protest and collective violence that it became known as "California's dirty little civil war." At first, Walton tried to use concepts that centered on social class and class conflict to understand this struggle. These were his initial, sensitizing concepts. He found that these concepts did not help him make sense of the evidence that he collected, nor did they direct him down data collection paths that advanced the study. Eventually he came to understand the struggle more in terms of collective responses anchored in local conditions to changing governmental structures, especially the growing influence and power of the federal government. These new concepts directed him to important historical evidence that he might have overlooked otherwise.

Clarifying Concepts and Categories

Qualitative research clarifies concepts (the key components of analytic frames) and empirical categories (which group similar instances of social phenomena) in a reciprocal manner. These two activities, categorizing and conceptualizing, go hand in hand because concepts define categories and the members of a category exemplify or illustrate the concepts that unite them into a category.

Generally, the members of a category are expected to be relatively homogeneous with respect to the concepts they exemplify. If a researcher found that only some flight attendants engage in emotion work, then it would be wrong to use the concept to characterize flight attendants. Suppose a researcher studying flight attendants found that only those flight attendants hired after a specific point in time engage in a lot of emotion work. It might be possible to trace this to a change in the training of flight attendants and perhaps to a conscious attempt by management to alter how flight attendants interact with passengers. The lack of fit between the concept "emotion work" and the broad category "all flight attendants" in this event would enrich the study, making it possible to narrow the relevant category to a subset of flight attendants—those subjected to a specific kind of training—and showing a direct connection to management intervention.

This example shows the importance of examining the members of a category to make sure that they all display the concepts they are thought to exemplify. Researchers develop concepts from the images that emerge from the categories of phenomena they examine. They then test the limits of the concepts they develop by closely examining the members of relevant categories. In the example just presented, the concept of emotion work emerged from images of flight attendants constructed by the investigator. Subsequent examination of all flight attendants—to see if they all engage in emotion work—would establish the limits of the relevant category.

Consider a second example of the interaction of categories and concepts, Howard Becker's (1953) early study of becoming a marijuana user. Becker studied several marijuana users and found that each went through a process of *learning* to become a user—of learning *how* to enjoy marijuana. This led him to speculate that all *marijuana users* (the category) go through a *social process of learning* (the concept) to enjoy marijuana. He elaborated the key steps in the process of becoming a user by interviewing more than fifty users in the Chicago area in the early 1950s. He found that most, more or less, went through the same process of learning how to enjoy marijuana.

However, Becker did encounter a few users who did not go through this process, and, although they were users, they said that they did not enjoy the drug. Becker described them as people who used marijuana for the sake of appearance—in order to appear to be a certain kind of person or to "fit in" with the people around them. Did this invalidate the idea that all users go through the same learning process? Becker solved the problem by narrowing the relevant category. He argued that the social process of learning how to enjoy marijuana applied only to those who used marijuana for pleasure, a category that embraced most, but not all, users. This narrowing made it possible for him to establish a closer correspondence between category (those who use marijuana for pleasure) and concept (the social process of learning how to use marijuana).

These examples show that the core issue in the clarification and elaboration of categories and concepts is the assessment of the degree to which the members of a category exemplify the relevant concept. Are the same elements present in each instance in more or less the same way? When encountering contradictory evidence (for example, flight attendants who don't do emotion work or marijuana users who did not go through the social process of learning how to enjoy marijuana), researchers have two choices. They can discard the concept they were developing and try to develop new ones—concepts that do a better job of uniting the members of the category. Or they can narrow the category of phenomena relevant to their concept and try to achieve a better fit with the concept.

Elaborating Analytic Frames

Because categories and concepts are clarified in the course of qualitative research, the researcher may not be certain what the research subject is a "case of" until all the evidence is collected and studied. Deciding that the research subject is a case of something and then representing it that way is often the very last phase of qualitative research.

The open character of qualitative research can be seen clearly in the role played by analytic frames in this strategy. In some research strategies (for example, quantitative research; see Chapter 6), the main purpose of the analytic frame is to express the theory to be tested in terms of the relevant cases and variables. In qualitative research, by contrast, there is often only a tentative, vaguely formulated analytic frame at the outset because it is developed in the course of the research.

As more is learned about the cases and as categories and concepts are clarified, the researcher can address basic questions: What is this case a case of? What are its relevant features? What makes the chosen research subject or site valuable, interesting, or significant? As qualitative researchers elaborate analytic frames, they also deepen their understanding of their cases. To describe the work of flight attendants as a case of emotion work (Hochschild 1983) suggests that there are other jobs that also require emotion work (for example, tour guides, camp counselors, waitresses, and so on) and that the emotion-work frame developed in the study of flight attendants may be applied to these other people-oriented service occupations.

Not all qualitative researchers develop analytic frames. Sometimes they leave this task to other researchers studying related cases. The development of analytic frames is challenging because it requires the extension of the concepts elaborated in one case to other cases. Many qualitative researchers are content to report detailed treatments of the cases they study and leave their analytic frames implicit and unstated. They feel that their cases speak well enough for themselves.

This unwillingness to generalize is found in all types of qualitative research, from observations of small groups to historical interpretations of the international system. For this reason, qualitative researchers are often accused of being "merely descriptive" and not "scientific" in their research. As should be clear by now, however, the process of representing research subjects is heavily dependent on the interaction between concepts and images, regardless of whether this interaction is made explicit by researchers when they represent their subjects. Without concepts, it is impossible to select evidence, arrange facts, or make sense of the infinite amount of information that can be gleaned from a single case. Like other forms of social research, qualitative research culminates in theoretically structured representations of social life—representations that reflect the regimen of social research.

Using Qualitative Methods

There are many textbooks on qualitative methods, and they describe qualitative methods in a variety of ways (see for example Denzin 1970, 1978; Glaser and Strauss 1967; McCall and Simmons 1969; Strauss 1987; Schwartz and Jacobs 1979). In part, this diversity of views follows from the emphasis on *in-depth* investigation and the fact that there are many different ways to achieve in-depth knowledge. In sociology, anthropology, and most other social sciences, qualitative methods are often identified with participant observation, in-depth interviewing, fieldwork, and ethnographic study. These methods emphasize the immersion of the researcher in a research setting and the effort to uncover the meaning and significance of social phenomena for people in those settings. These techniques are best for studying social situations at the level of person-to-person interaction.

For an anthropologist, this immersion might involve living in some isolated village in some faraway part of the world. Consider, for example Margaret Mead's work *Coming of Age in Samoa* (1961). For a sociologist, immersion might involve long periods of observing and talking to people in one setting, such as Erving Goffman's research on the staff and patients of a mental institution, reported in his classic study *Asylums* (1961). In both examples, the organizing principle of the research is the idea that the kind of in-depth knowledge needed for a proper representation of the research subject must be based on the perspectives of the people being studied—that their lives and their worlds must be understood "through their eyes." In short, the emphasis is on immersion and empirical intimacy (Truzzi 1974).

The goal of this presentation of qualitative methods, however, is to address procedures that are relevant to all types of qualitative research, not simply the work of those who seek to represent social life and as it appears through the eyes of participants. Researchers who seek to represent historically significant events, for example, cannot hope to see these events through the eyes of the participants if these events occurred in the distant past (the French Revolution, for example, or slavery in the U.S. South). Still, these historical researchers, like others who use qualitative methods, value and seek in-depth knowledge about cases, and they attempt to piece together meaningful images from evidence, with the help of concepts and analytic frames.

The key features common to all qualitative methods can be seen when they are contrasted with quantitative methods. Most quantitative data techniques are *data condensers*. They condense data in order to reveal the big picture. For example, calculating the percentage of unionized workers who vote for the Democratic party condenses information on thousands of individuals into a single number showing the link between these two attributes (union membership and party preference). Qualitative methods, by contrast, are best understood as *data enhancers*. When data are enhanced, it is possible to see key aspects of cases more clearly, depending on how it is done.

In many ways, data enhancement is like photographic enhancement. When a photograph is enhanced, it is possible to see certain aspects of the photographer's subject more clearly. When qualitative methods are used to enhance social data, researchers see things about their subjects that they might miss otherwise. Data enhancement is the key to in-depth knowledge.

Almost all qualitative research seeks to construct representations based on in-depth, detailed knowledge of cases, often to correct misrepresentations or to offer new representations of the research subject. Thus, qualitative researchers share an interest in procedures that clarify key aspects of research subjects—procedures that make it possible to see aspects of cases that might otherwise be missed. While there are many such procedures, two that are common to most qualitative work are emphasized here: analytic induction and theoretical sampling. Both techniques are data enhancers.

Analytic Induction

Analytic induction means very different things to different researchers. Originally, it had a very strict meaning and was identified with the search for "universals" in social life (Lindesmith 1947; Cressey 1953; Turner 1953; Robinson 1951). Universals are properties that are invariant. If all upper middle-class white males over the age of fifty in the United States voted for the Republican party, then this would constitute a "universal." If only one person in this category voted for some other party, then the pattern would not be universal and thus would not qualify as a finding, according to a very strict, very simple-minded application of the method of analytic induction. Today, however, analytic induction is often used to refer to any systematic examination of similarities that seeks to develop concepts or ideas.

Rather than seeing analytic induction as a search for universals, a search that is likely to fail, it is better to see it as a research strategy that directs investigators to pay close attention to evidence that challenges or disconfirms whatever images they are developing. As researchers accumulate evidence, they compare incidents or cases that appear to be in the same general category with each other. These comparisons establish similarities and differences among incidents and thus help to define categories and concepts. (Sociologists Barney Glaser and Anselm Strauss call this process the **constant comparative method**.) Evidence that challenges or refutes images that the researcher is constructing from evidence provides important clues for how to alter concepts or shift categories.

A study in a hospital might examine the care given to dying patients. By comparing cases of this type, the researcher can identify common features and the major dimensions of variation among incidents. Based on hours of observing the care of dying patients, a researcher might find:

1. that nurses and other hospital personnel implicitly evaluate the potential "social loss" represented by each patient if the patient were to die
2. that a small number of patient characteristics enter into this evaluation (for example, the age and education of the patient)
3. that the quality of patient care depends on the potential social loss inferred by the hospital personnel

Incidents that challenge either the generality of the evaluation of the social loss of dying patients or the impact of this evaluation on the care patients receive would be especially important for refining these ideas. In the next phase of the research, the investigator might seek out disconfirming evidence (for example, a patient who is judged to be not much of a "social loss" but nevertheless receives excellent care) to test these initial images and see how they need to be revised or limited. If, for example, the researcher found that hospital personnel ignored the social loss represented by accident victims, then he or she would be forced either to reformulate the image to accommodate accident victims or else limit its applicability to nonaccident patients.

In effect, the method of analytic induction is used both to construct images and to seek out contrary evidence because it sees such evidence as the best raw material for improving initial images. As a data procedure, this technique is less concerned with how much positive evidence has been accumulated (for example, how many cases corroborate the image the researcher is developing), and more with the degree to which the image of the research subject has been refined, sharpened, and elaborated in response to both confirming and disconfirming evidence.

Analytic induction facilitates the reciprocal clarification of concepts and categories, a key feature of qualitative research. When Howard Becker narrowed his category from "all marijuana users" to "those who use marijuana for pleasure," he used the technique of analytic induction. Essentially, the technique involves looking for relevant similarities among the instances of a category, and then linking these to refine an image (for example, the image of how one becomes a marijuana user). If relevant similarities cannot be identified, then either the category is too wide and heterogeneous and should be narrowed, or else the researcher needs to take another look at the evidence and reconceptualize possible similarities. Negative cases are especially important because they are either excluded when the relevant category is narrowed, or they are the main focus when the investigator attempts to reconceptualize commonalities and thereby reconcile contradictory evidence.

Consider a more detailed example: Jack Katz (1982) studied legal assistance lawyers—those who help poor people. He found that many legal assistance lawyers burn out quickly—in less than two years—and abandon this kind of work, often for more lucrative legal careers. Katz wanted to understand why by studying those who stayed with legal assistance work despite its drawbacks. He assembled evidence on the legal assistance lawyers in the group he studied and checked out several of his initial ideas by comparing those who had quit before two years of service with those who had stayed on for more than two years.

One of the first ideas Katz examined was based on his initial impressions of these attorneys. He speculated that legal assistance lawyers who were former political activists did not burn out like the others. A systematic examination of the evidence on many lawyers provided some support for this speculation. However, the fit was far from perfect. There were some who stayed with legal assistance work who were not former political activists, and there were former political activists who left legal assistance work before two years had elapsed.

Katz examined these negative cases closely and found some problems with his initial formulation. Some former activists left for obvious reasons. They were offered positions that were clearly a step up, careerwise. Some who were not former activists stayed because they lacked alternatives—they couldn't get better jobs as lawyers—or because they had positions in the organization that they liked (such as administrative positions).

It was clear to Katz that his categories "staying versus leaving" had to be refined and that his search for adequate explanatory concepts was far from over. First, he narrowed the category that interested him most—those who stayed. Clearly he was not interested in all stayers. Some stayers, after all, had interesting work within the legal assistance organization he studied. Rather, he was interested in people who stayed despite being involved in frustrating or limiting work. He restricted his focus to this subset of stayers and searched for relevant similarities within this group.

With this shift he became less interested in all stayers versus all leavers and more interested in differences between categories of stayers—those who stayed despite frustrating work versus other stayers. In short, the focus was on *how* people stayed, and he had straightforward explanations for many stayers (for example, those with interesting work). As it turned out, this tighter category—stayers with frustrating work—also proved to be too broad, and he later narrowed it further to legal assistance lawyers who were involved in low-status work. After all, some lawyers doing significant work, he discovered, were nevertheless frustrated with their work.

The search for explanatory factors became more focused as the main category of interest narrowed. After rejecting "activist background" as an explanation for staying, Katz tried to distinguish lawyers who were more oriented toward using the legal system for reform from those who were less so. He also looked at the participation of lawyers in social activities that celebrated reform work (for example, progressive political groups). This search for important commonalities among stayers went hand in hand with narrowing the relevant category of stayers from all stayers to those who were involved in low-status work.

The process of narrowing and refining is depicted in Table 4.1, which shows the process of analytic induction in tabular form based on Jack Katz's description. The table reports hypothetical information on thirty lawyers to illustrate the general process he describes, not his specific conclusions. The first three columns show the narrowing of the category of stayers, from all stayers (column 1; 18 out of 30 lawyers) to stayers with frustrating work (column 2; 13 out of 30 lawyers), to stayers involved in work that carried low status (column 3; 10 out of 30 lawyers). Columns 4 through 6 show the various ways Katz tried to explain staying—his various images of the "stayer." As his focus shifted from column 1 to column 2 and then to column 3, he became more interested in how and why people stayed and less in the difference between stayers and the twelve leavers at the bottom of the table. In other words, he came to view staying as an accomplishment for those doing low-status work and studied how it was accomplished.

First, Katz tried to construct an image of staying as a continuation of a commitment to political activism (column 4). As the hypothetical data in Table 4.1 show, this image fails. Of the eighteen lawyers who stayed more than two years, only seven were former activists, and of the twelve who left the organization, four were former activists. Next, Katz studied his negative cases closely (especially, nonactivists who stayed) and found that his categorization of stayers versus leavers was too crude. He reasoned that what really interested him most was people who stayed despite their involvement in frustrating work. He then tried to find commonalities among this subset of stayers, looking at their reform orientations and their participation in a social life supportive of reform work. The fit was still not close enough. There were some lawyers who did frustrating work, for example, who were not reform oriented.

Examination of negative cases led to a further narrowing of the category—to lawyers involved in low-status work—and further refinement of the image—to participation in a social environment that glorified reform work. These further refinements resulted in a good fit. The data in the table suggest that legal aid lawyers will do low-status work if they participate in a social environment that glorifies the idea that important social reforms can be achieved through the legal system.

TABLE 4.1

Hypothetical Example of Analytic Induction

	Categories			*Explanatory Concepts*		
	1	*2*	*3*	*4*	*5*	*6*
Case	*Stayed More Than Two Years?*	*Works in a Frustrating Place?*	*Involved in Low-Status Work?*	*Activist Background?*	*Reform Oriented?*	*Social Life Supports Reform Orientation?*
1	yes	yes	yes	yes	yes	yes
2	yes	yes	yes	yes	yes	yes
3	yes	yes	yes	yes	yes	yes
4	yes	yes	yes	yes	yes	yes
5	yes	yes	yes	yes	yes	yes
6	yes	yes	yes	yes	yes	yes
7	yes	yes	yes	yes	yes	yes
8	yes	yes	yes	no	yes	yes
9	yes	yes	yes	no	yes	yes
10	yes	yes	yes	no	yes	yes
11	yes	yes	no	no	yes	no
12	yes	yes	no	no	no	no
13	yes	yes	no	no	no	no
14	yes	no	no	no	no	no
15	yes	no	no	no	no	no
16	yes	no	no	no	no	no
17	yes	no	no	no	no	no
18	yes	no	no	no	no	no
19	no	no	no	yes	no	no
20	no	no	no	yes	no	no
21	no	no	no	yes	no	no
22	no	no	no	yes	no	no
23	no	no	no	no	no	no
24	no	no	no	no	no	no
25	no	no	no	no	no	no
26	no	no	no	no	no	no
27	no	no	no	no	no	no
28	no	no	no	no	no	no
29	no	no	no	no	no	no
30	no	no	no	no	no	no

Columns 3 and 6 correspond perfectly. In fact, most qualitative researchers are satisfied with less than perfect fit. There is usually at least a handful of extraneous evidence that neither fits nor challenges a particular image. The goal is not perfect fit, per se, but a conceptual refinement that provides a deeper understanding of the research subject. Basically, the greater the effort to account for or understand negative cases or contrary evidence, the deeper the understanding of the research subject. The technique of analytic induction thus facilitates the goal of in-depth knowledge.

Katz comments that analytic induction is poorly labeled because it is not a technique of pure induction. Researchers work back and forth between their ideas and their evidence, trying to achieve what Katz calls a "double fitting" of explanations and observations (that is, ideas and evidence). As discussed in Chapter 3, this process of double fitting is best understood as retroduction, a term that describes the interplay of induction and deduction in the process of scientific discovery.

Theoretical Sampling

Sometimes qualitative researchers conduct investigations of related phenomena in several different settings. Most often this interest in a broader investigation follows from a deliberate strategy of **theoretical sampling**, a term coined by Barney Glaser and Anselm Strauss (1967) to describe the process of choosing new research sites or cases to compare with one that has already been studied. For example, a researcher interested in how environmental activists in the United States maintain their political commitments might extend the study to (1) environmental activists in another part of the world (for example, Eastern Europe) or perhaps to (2) another type of activist (for example religious activists in the United States).

This process of theoretical sampling occurs not only in the study of social groups (for example, environmental activists), but also in the study of historical processes and episodes. General questions that arise in a study of the Russian Revolution of 1917 might be addressed by examining the Chinese Revolution of 1949 or the recent Nicaraguan Revolution. There may be questions about the role of peasants in the Russian Revolution that could be answered by examining the Chinese case and comparing it to the Russian case.

The choice of the comparison group (comparing environmental activists in the United States with either environmental activists in Eastern Europe or with people in the United States who maintain radical religious commitments) can vary widely depending on the nature and goals of the investigation. Different comparisons hold different aspects of cases constant. Comparing environmental and religious activists in the United States holds some things constant such as the impact of national setting, but allows the nature of the commitment to vary (environmental versus religious). Comparing environmental activists in the United States with environmental activists in Eastern Europe highlights the impact of the factor that varies most, national setting, but holds the nature of the commitment, environmental, constant.

When a researcher employs a strategy of theoretical sampling, the selection of additional cases is most often determined by questions and issues raised in the first case studied. Selection of new cases is not a matter of convenience; the researcher's sampling strategy evolves as his or her understanding of the research subject and the concepts it exemplifies matures. The goal of theoretical sampling is not to sample in a way that captures all possible variations, rather in one that aids the development of concepts and deepens the understanding of research subjects.

A researcher studying how hospital personnel evaluate the potential social loss of dying patients and link the care they give to these evaluations might believe that this practice is caused by limited resources in the hospital studied. If the hospital had more resources (for example, more nurses), it might be able to provide better and more uniform care to all patients, regardless of their social value. To explore this idea, the researcher might study two additional hospitals, one with more resources and one with fewer resources than the first hospital. If the reasoning based on the first hospital is correct, then the staff of the hospital with more resources should spend less time evaluating the social loss of dying patients and provide more uniform care, while the staff of the hospital with fewer resources should spend more time evaluating social loss and should adjust their care in more strict accordance with these evaluations.

This expansion of the study to two new sites is a straightforward implementation of the idea of theoretical sampling. The selection of the new sites follows directly from ideas developed in the first site and provides an opportunity to confirm and deepen the insights developed in that setting. Of course, if research in these new settings were to contradict expectations based on research in the first hospital, then the researcher would be compelled to develop a different understanding of how and why hospital personnel varied their care of dying patients.

This example of theoretical sampling also shows that it is a technique of **data triangulation** (Denzin 1978). Triangulation is a term that originally described how sailors use stars and simple trigonometry to locate their position on earth. More generally, triangulation can be understood as a way of using independent pieces of information to get a better fix on something that is only partially known or understood. In the example just presented, the researcher used evidence from two other hospitals, one with more resources and one with fewer, to get a better fix on the first hospital. By comparing the three hospitals, arrayed along a single continuum of resources, the researcher could assess the validity and generality of findings from the first hospital.

Theoretical sampling is also a powerful technique for building analytic frames. Helen Rose Fuchs Ebaugh (1977) studied ex-nuns—women who left Catholic religious orders—and used this group of women to develop the concept of "role exit," in much the same way that Arlie Hochschild used her study of flight attendants to develop the concept of emotion work. Ebaugh became interested in people whose current self-identities were strongly influenced by the roles they had left behind. This interest led her to develop a deliberate strategy of sampling different kinds of "exs" in addition to ex-nuns: ex-doctors, mother's without custody, transsexuals, and so on. Each group offered evidence on a different type of role exit, the most dramatic being an exit from one sex to another. The end product of Ebaugh's strategy of theoretical sampling was a fully developed analytic frame for role exit (Ebaugh 1988).

Howard Becker (1963) studied a variety of groups classified as "deviant" in addition to marijuana users. He joined these different cases together in a single analytic frame and called all these groups "outsiders." His frame emphasized a dual process of *social learning* (people learn "deviant" behaviors from others in social settings) and *labeling* (society's tendency to label some groups deviant furthers their isolation from the larger society). His work challenged conventional thinking that certain types of people were at a greater risk of becoming deviant and focused subsequent research on social processes. In a similar manner, Erving Goffman (1963) studied a wide variety of stigmatized people, from those with physical handicaps to homosexuals. From a consideration of many different types, he developed a powerful analytic frame for understanding how stigmatized individuals deal with their discredited identities.

While the strategy of theoretical sampling is an excellent device for gaining a deeper understanding of cases and for advancing theory (one of the main goals of social research), many qualitative researchers consider the representation of even a single case sufficient for their goals. Some consider the addition of new cases—using the strategy of theoretical sampling—to be a useless detour from the important task of understanding one case well. They are content to leave the comparison of cases and the development of broad analytic frames to researchers more interested in general questions.

While this reluctance to broaden an investigation is common among qualitative researchers, the strategy of theoretical sampling offers a powerful research tool. As Glaser and Strauss (1967) argue, theoretical sampling offers the opportunity to construct generalizations and to deepen understanding of research subjects at the same time.

The Study of a Single Case

The techniques of analytic induction and theoretical sampling work best when there are multiple instances of the phenomenon the researcher is studying. The study of the care of dying patients, for instance, involves observing how patients are treated. Each patient provides another instance to examine. What techniques can researchers use when they study only a single instance—for example, one person's life or a single historical event? While it is true that most data procedures are designed for multiple instances, the study of a single case is not haphazard and unstructured (Feagin et al. 1991). In fact, the single-case study is structured in ways that parallel analytic induction.

For illustration, consider a researcher who seeks to evaluate the historical significance of the resignation of President Richard Nixon, who left office in the middle of his second term. Suppose the goal of the researcher in this investigation is to try to interpret this episode as a serious blow to the authority of the U.S. government, at least in the eyes of the American people. Because of what transpired, according to this interpretation, the American people could never again see their politicians as statesmen or trust government leaders and officials to tell them the truth.

Of course, there are many different ways to interpret each historical episode, and each interpretation is anchored in a different analytic frame. The researcher's interpretation sees the events surrounding the resignation of President Nixon in terms of the authority and legitimacy of governments. What kinds of conditions and events enhance a government's authority? What kinds undermine its authority?

In order to evaluate this interpretation, the researcher would have to assemble facts relevant to the analytic frame (which emphasizes factors influencing a government's authority) and see if they can be assembled into an image that supports this interpretation. Of course, there are many facts, and not all will necessarily be consistent with the initial interpretation. The key question is: among the relevant facts, which are consistent and which are not? Analytic frames play an important part in this process because they define some facts as relevant and others as irrelevant, and different frames define different sets of facts as relevant.

In many ways, this evaluation of facts is like analytic induction. In analytic induction the goal is to see if all the relevant instances are the same with respect to some cause or characteristic, as in Jack Katz's research on legal assistance attorneys. In the study of a single case, the problem is to see if all the facts that are relevant in some way to the suggested frame agree with or support an interpretation. Thus, the different facts in the study of a single case are like the different instances in analytic induction.

Often the facts relevant to a particular frame, once assembled, do not provide strong support for the initial interpretation. As in analytic induction, the interpretation and the facts are "double fitted." That is, there is an interplay between the researcher's interpretation and the facts, an interaction that moves either toward some sort of fit or toward a stalemate. As in the study of many instances (for example, the care of many different patients in a hospital), the interplay between evidence-based images and theoretical ideas expressed through analytic frames leads to a progressive refinement of both.

It is important to remember that each different interpretation is anchored in a different frame. Thus, the facts relevant to one frame will not overlap perfectly with the facts relevant to another. Thus, there can be many different ways to frame a single case, and each interpretation may be valid because of this imperfect overlap. Cases that can be interpreted in a variety of different ways are considered "rich" because they help researchers explore the interconnection of the ideas expressed through different frames.

Conclusion

Researchers use qualitative methods when they believe that the best way to construct a proper representation is through in-depth study of phenomena. Often they address phenomena that they believe have been seriously misrepresented, sometimes by social researchers using other approaches, or perhaps not represented at all. This in-depth investigation often focuses on a primary case, on the commonalities among separate instances of the same phenomenon, or on parallel phenomena identified through a deliberate strategy of theoretical sampling.

Qualitative methods are holistic, meaning that aspects of cases are viewed in the context of the whole case, and researchers often must triangulate information about a number of cases in order to make sense of one case. Qualitative methods are used to uncover essential features of a case and then illuminate key relationships among these features. Often, a qualitative researcher will argue that his or her cases *exemplify* one or more key theoretical processes or categories. Finally, as qualitative research progresses, there is a reciprocal clarification of the underlying character of the phenomena under investigation and the theoretical concepts that they are believed to exemplify.

5

Using Comparative Methods to Study Diversity

Introduction

Comparative researchers examine patterns of similarities and differences across a moderate number of cases. The typical comparative study has anywhere from a handful to fifty or more cases. The number of cases is limited because one of the concerns of comparative research is to establish familiarity with each case included in a study. Like qualitative researchers, comparative researchers consider how the different parts of each case—those aspects that are relevant to the investigation—fit together; they try to make sense of each case. Thus, knowledge of cases is considered an important goal of comparative research, independent of any other goal.

While there are many types of comparative research (see Przeworski and Teune 1970; Skocpol 1984; Tilly 1984; Stinchcombe 1978; Lijphart 1971), the distinctiveness of the comparative approach is clearest in studies that focus on diversity (Ragin 1987). Recall that the qualitative approaches examined in Chapter 4 emphasize commonalities, and the primary focus is on similarities across instances (for example, the fact that hospital personnel assess the potential social loss of each dying patient). This concern for commonalities dovetails with an interest in clarifying categories and concepts (for example, the concept of potential social loss and the situations in which it is assessed). In comparative research on diversity, by contrast, the category of phenomena that the investigator is studying is usually specified at the outset, and the goal of the investigation is to explain the diversity within a particular set of cases (see, for example, Lijphart 1984; Rueschemeyer et al. 1992; Moore 1966; Nichols 1986). (This type of comparative research, which is the major focus of this chapter, is examined in detail in Ragin 1987.)

Consider the following example of comparative research on diversity. From the middle 1970s through the 1980s many less developed countries experienced mass protest in response to austerity programs demanded by the International Monetary Fund (IMF). These countries had accumulated large public debts that they could not repay. In exchange for better terms (for example, lower interest rates and longer repayment periods), the governments of these countries agreed to IMF mandates that they implement economic policies designed to facilitate debt repayment. For instance, in some countries the IMF demanded that the government stop subsidizing the prices of basic commodities such as fuel and food. These austerity measures saved governments money and made debt repayment more feasible; they also provoked widespread protest among citizens faced with new challenges to their economic well-being (Walton and Ragin 1990).

A comparative researcher interested in these countries might contrast the different *forms* of protest that occurred in response to these austerity programs. In some countries, there were riots; in others, there were labor strikes led by unions; in others, there were mass demonstrations involving many different groups; in others, opposition political parties organized protests; and so on. Why did different kinds of protest erupt in different countries? What causal conditions explain these different responses to austerity programs? And why did some countries with severe austerity programs experience very mild mass protest?

To explain this diversity, a comparative researcher would first group countries according to their different responses to austerity, placing all those with riots in one category, those with demonstrations in another, and so on. Next, the investigator would look for patterns of similarities and differences. What are the similarities among the countries with riots that distinguish them from all other countries? Perhaps, the countries with riots also had repressive governments, widespread poverty, and serious crowding in major urban areas, and none of the nonriot countries had this specific combination of conditions. How did the countries with mass demonstrations differ from all the other countries with austerity programs? This search for systematic differences would continue until the researcher could account for the diverse responses to austerity found in these countries.

Thus, in research that emphasizes diversity the focus is on the similarities within a category of cases with the same outcome (for example, countries with riots) that (1) distinguish that category from other categories (countries with other forms of austerity protest) and (2) explain the outcome manifested by that category. In other words, the study of diversity is the study of patterns of similarities and differences within a given set of cases (in this example, countries with austerity protests).

Contrasts with Other Research Strategies

As noted already, the main difference between comparative research on diversity and qualitative research on commonalities is that their basic orientation toward cases differs. When qualitative researchers study commonalities they usually view multiple cases as many instances of the same thing. A qualitative researcher who interviews many taxicab drivers uses these many instances to deepen the portrayal of this case—the taxicab driver.

Comparative researchers who study diversity, by contrast, tend to look for differences among their cases. Comparative researchers examine patterns of similarities and differences across cases and try to come to terms with their diversity. A comparative researcher might study the settlement of Indochinese refugees in the United States in the 1970s and 1980s, contrasting the ways they were received in a variety of different communities. It might be possible to distinguish four or five basic types of receptions—from hostile to indifferent to open to paternalistic and so on—and then to pinpoint the factors (such as size and wealth of community) that determined these different receptions.

Another comparative researcher might study bars in a community and contrast the different strategies they use to encourage and discourage drinking. Bars that cater to different customers (for example, bikers versus business people versus lesbians) surely use different techniques. In each of these examples, the research focuses on the diversity that exists within a specific set of cases.

Quantitative researchers (the focus of Chapter 6) also examine differences among cases, but with a different emphasis. In quantitative research, the goal is to explain the covariation of one variable with another, usually across many, many cases. A quantitative researcher might explain different levels of income across thousands of individuals included in a survey by pointing to the covariation between income levels and educational levels—people with more education tend to have more income. In quantitative research, the focus is on differences in levels and how different variables like income and education covary across cases. In comparative research, by contrast, the focus is on diversity—*patterns* of similarities and differences.

Furthermore, the quantitative researcher typically has only broad familiarity with the cases included in a study. As the number of cases exceeds fifty or so, it becomes increasingly difficult to establish familiarity with each case. Imagine a survey researcher trying to become familiar with the lives of the thousands of people included in a telephone survey or a political scientist trying to keep up with major events in all countries. Neither task is feasible. There are practical limits to how many cases a researcher can study closely.

The Goals of Comparative Research

The emphases of comparative research on diversity (especially, the different patterns that may exist within a specific set of cases) and on familiarity with each case make this approach especially well suited for the goals of exploring diversity, interpreting cultural or historical significance, and advancing theory.

Exploring Diversity

The comparative approach is better suited for addressing patterns of diversity than either of the other two strategies. Diversity is most often understood in terms of types of cases. The typical goal of a comparative study is to unravel the different causal conditions connected to different outcomes—causal patterns that separate cases into different subgroups. This explicit focus on diversity distinguishes the comparative approach from the qualitative approach. Recall that in qualitative research the goal is often to clarify categories with respect to the concepts they exemplify by examining similarities across the instances of a category (such as taxi drivers).

One common outcome of comparative research is the finding that cases that may have been defined as "the same" at the outset are differentiated into two or more categories at the conclusion of the study. For example, a researcher studying major U.S. cities that have elected African-American mayors might conclude at the end of the study that there are two major types of cities—those where interracial alliances elected African-American mayors and those where black voters, who happened also to constitute a majority of voters, elected African-American mayors. The political dynamics and significance of the elections could differ considerably across the two types.

The researcher studying governments that terrorize citizens who oppose them might find that there are several main types, depending on the international standing of the government in question. For instance, when a government is supported by the United States and other major powers, its terror may be overlooked. When a government lacks this support, its terror may be considered repugnant. Governments in the second category would have to contend with the possibility that their actions might provoke international sanctions or intervention and therefore practice more covert forms of terror.

While comparative researchers often discern types in the course of their examinations of patterns of diversity, they may also begin their research with a tentative delineation of types. A common strategy is to categorize cases according to their different outcomes. The goal of the research in this case is to unravel the causal conditions that generate different outcomes. If different causes can be matched to the different outcomes, then the research confirms the investigator's understanding of the factors that distinguish these cases. If not, then the frame for the research needs to be reformulated.

For example, a researcher might examine the causes of different types of government repression. Some repressive governments, for example, may simply harass their opponents—incarcerating them for short periods, subjecting them to frequent questioning, opening their mail, and so on. Other repressive governments may torture and kill their opponents. Still other governments may focus their repressive energies not on opponents, but on purging the less committed from their own ranks—members of the ruling political party or clique. Still others may attack random members of society in order to maintain a general state of terror and obedience (as Stalin did in the Soviet Union). It is important to understand different types of repressiveness and the various conditions that explain the emergence of each type.

The goal of exploring diversity is important because people, social researchers included, sometimes have trouble seeing the trees for the forest. They tend to assume uniformity or generality when, in fact, there is a great deal of diversity. A simple example: generally, governments that are less democratic tend to be more repressive. However, there are many instances of repression by democratically elected governments and many instances of politically tolerant and lenient governments that are not democratic. To understand government repression fully, it is necessary to go beyond the simple identification of political repression with an absence of democracy and examine the different forms of government repression that exist in all countries.

Interpreting Cultural or Historical Significance

Comparative researchers focus explicitly on patterns of similarities and differences across a range of cases. Relevant cases, in turn, are almost always drawn from a specific and known set. Recall that in qualitative research (Chapter 4), much energy is often devoted to coming to terms with the case. What is this case a case of? What concepts are exemplified in this case? Into which larger social scientific categories, if any, does it fit? In comparative research, by contrast, the researcher usually starts with a good sense of the larger category that embraces the cases included in the study because this category is usually specified beforehand (such as "countries with austerity protests").

A researcher might focus on "military coups in Latin America since 1975" or "major cities in the United States that have elected African-American mayors" or "recent U.S. federal court cases involving the rights of AIDS patients" (Musheno et al. 1991). In each example, the relevant set of cases is defined in advance, and there is a finite, usually moderate number of such cases. Typically, the category that establishes the boundary of the set is historically and geographically delimited. In each of these examples time and place boundaries are either plainly stated (for example, "Latin America since 1975") or implied ("recent U.S.").

This focus on circumscribed categories makes the comparative strategy well suited for the goal of interpreting historically or culturally significant phenomena, especially when there is a moderate number of cases, as in the examples just mentioned. The category "major cities in the United States that have elected African-American mayors" is historically significant in part simply because it is a relatively new and major phenomenon. Prior to the expansion of civil rights in the 1960s, there were no African-American mayors in major U.S. cities. It is culturally significant because of the relevance of race and race relations to American society. Likewise, the category "military coups in Latin America since 1975" is significant to those concerned with progress of democracy and human rights in this region.

Because the comparative approach focuses on differences between cases and the differentiation of types, it facilitates historical interpretation. Consider the category *revolution*. Some revolutions simply change those who are in power or alter other political arrangements without implementing any major changes in society. The revolutionaries that overthrew Ferdinand Marcos in the Philippines did not attempt any fundamental changes in Philippine society. Other revolutions, by contrast, bring with them regimes that seek to alter society fundamentally. Kings are beheaded; property is confiscated; basic social patterns and relations are changed forever. Revolutionary social changes of this nature were attempted after the French Revolution of 1789, the Russian Revolution of 1917, and the Chinese Revolution of 1949.

Revolutions that attempt fundamental social change are treated as a distinct type by social scientists. These massive upheavals of society are called *social revolutions* to distinguish them from revolutions that simply change leaders or other political arrangements (Skocpol 1979). By differentiating social revolutions from all other forms, researchers provide important tools for understanding and interpreting these massive social transformations. When a major upheaval occurs, researchers can assess whether or not it qualifies as a social revolution. If so, it can be compared with other social revolutions. If not, then some other category may be used (for example, *coup d'etat*) to interpret the event and to specify comparable cases. Generally, when a set of comparable cases can be specified, these cases aid the interpretation and understanding of the new case.

More generally, when social scientists categorize an event, they establish a primary analytic frame for its interpretation. Thus, the interpretation of historically or culturally significant events is often a struggle over the proper classification of events into broad categories—a key concern of the comparative approach.

Advancing Theory

Several basic features of the comparative approach make it a good strategy for advancing theory. These features include its use of flexible frames, its explicit focus on the causes of diversity, and its emphasis on the systematic analysis of similarities and differences in the effort to specify how diversity is patterned.

In comparative research, investigators usually initiate research with a specific analytic frame, but these initial frames are open to revision. The researcher interested in military coups in Latin American since 1975 already has a frame for the research—the frame for military coups. Recall that the frames of qualitative research are fluid, and researchers may not finish developing their frames until after all the work of collecting and studying the evidence is complete. In comparative research, by contrast, frames are established at the outset of a research project, but they remain flexible. Comparative researchers expect their frames to be revised, and in fact conduct research in order to sharpen the ideas expressed in a frame.

A researcher interested in welfare states in advanced countries might start out with a frame that specifies two basic types of welfare states but then conclude with a specification of three or four types (Esping-Andersen 1990). Or, the researcher might conclude that there is only one main type and that all deviations from this main type are best understood as underdeveloped or incomplete expressions of the main type (J. Stephens 1979). By altering initial frames in response to evidence, comparative researchers refine and elaborate existing ideas and theoretical perspectives.

When conducting their research, comparative researchers are more explicitly concerned with causation and causal complexity than are most qualitative researchers. For example, when comparative researchers differentiate types (such as types of government repression), they also try to specify the combinations of causal conditions conducive to each type. What causes some regimes to concentrate their repressive efforts on regime opponents? What causes others to focus their efforts on purging troublesome members of the ruling party? And what causes still other regimes to cultivate a general state of terror in the population at large? This emphasis on causation is central to theory because most theories in the social sciences are concerned with explaining how and why—that is, with specifying the causes of social phenomena.

To assess causation, comparative researchers study how diversity is patterned. They compare cases with each other and highlight the contrasting effects of different causes. Comparative researchers view each case as a combination of characteristics (for example, conditions relevant to government repression) and examine similarities and differences in combinations of characteristics across cases in their effort to find patterns.

The Process of Comparative Research

The comparative study of diversity is neither as fluid as qualitative research nor as fixed as quantitative research. Comparative researchers typically start with a carefully specified category of phenomena that is intrinsically interesting in some way (for example, countries with austerity protests). They use analytic frames to help them make sense of their categories, and they revise their frames based on their examination of evidence.

In the course of their research they focus on patterns of similarities and differences among cases and assess patterns of diversity. This assessment of diversity provides the foundation for improving or revising the analytic frame chosen at the outset of the study. Like qualitative research, the comparative approach stimulates a rich dialogue between ideas and evidence. Researchers generate images from their data and adjust their frames as they construct representations of their research subjects.

Selecting Cases

Comparative researchers usually initiate their research with a specific set of cases in mind. Most often, this set has clear spatial and temporal boundaries and embraces cases that are thought to be comparable with each other, as in the examples already described. The degree to which the cases that are selected actually belong to the same category (and therefore are comparable) is assessed in the course of the research. While conducting the investigation, the researcher may decide that some cases don't belong in the same category as the others and can't be compared. They also may reformulate the category as the research proceeds. Usually, however, such adjustments are modest.

Typically, the cases that comparative researchers select for study are specific to their interests and to those of their intended audience, for example, countries with mass protest against austerity programs. This category of countries has clear spatial and temporal boundaries and embraces a set of comparable cases. It is also an intrinsically interesting set of cases. In short, it is just the kind of delimited empirical category that is well suited for comparative investigation.

The comparative approach can be applied to many different kinds of cases, not just countries. It is important, however, for the cases selected to be comparable and to share membership in a meaningful, empirically defined category. For example, the comparative approach can be applied to the fraternities on a college campus, to ethnic and racial groups living in a major urban area, to different religious congregations in a medium-size town, to the truck stops along Interstate 55, or to the elections in the congressional districts of a large state. The set of cases must be coherent. Usually, they must also offer some potential for advancing social scientific thinking.

Using Analytic Frames

When researchers choose their cases, they also usually select their analytic frames. Essentially, a frame is chosen when the researcher specifies what about the cases is of interest. The researcher interested in countries with austerity protests may be interested in the different forms that the protest took. This frame, which would be developed from the existing social science literature on mass protest, would specify how people respond to different conditions in different ways when they engage in political protest. In short, it would detail the different kinds of factors the researcher should examine in a comparative study of mass protest.

In some countries opposition groups may have many resources; in others, they may have few. Groups with more resources are more likely to engage in organized activities such as strikes and in other activities that are relatively costly to participants. People on strike must give up their wages. Thus, this frame, which would be developed from the existing literature on social movements and collective action (for example, Jenkins 1983), would direct the researcher to focus on resources, among other things. Analytic frames help researchers see aspects of cases that they might otherwise overlook, and direct their attention away from other aspects.

Sometimes researchers are interested in many facets of their cases and don't select a frame until they are well along in their research. It might take a while, for example, to determine what a comparison of countries with austerity protests might best offer in the way of general social scientific knowledge. Comparative researchers also may develop new frames from their evidence, for example, a new frame for the study of race and politics based on a study of cities where coalitions of white and black voters have elected African-American mayors. This practice is less common in comparative research than in qualitative research, however, because comparative researchers start with a fairly good sense of their cases and the empirical category that embraces them (such as "countries that experienced mass protest in response to IMF-mandated austerity").

Analyzing Patterns of Diversity

In comparative research the examination of diversity—patterns of similarities and differences—goes hand in hand with the study of causes. Generally, researchers expect different causal conditions to be linked to divergent outcomes in interpretable ways. Thus, the goal of the researcher's examination of patterns of similarities and differences is to identify causal links—how different **configurations** of causes produce different outcomes across the range of cases included in a study. The specification of different patterns of causation is the primary basis for the differentiation of types.

In a study of how sororities generate a feeling of group solidarity, different ways of generating this feeling should affect the nature of the solidarity that is generated. The researcher might find that some sororities generate solidarity around special events and rituals, while others generate it through routine activities that bring members of the sorority together on a daily basis. These different ways of generating this feeling should have consequences for the nature of the solidarity observed. For example, solidarity in sororities of the first type may be more visible but also less durable, while in the second type, it may be more subtle but more enduring.

If causes and outcomes cannot be linked in interpretable ways, then researchers must reexamine their specification of causes and outcomes and their differentiation of types. In many ways this process of differentiating types and specifying causal links specific to each type resembles the "double fitting" of categories and images that constitutes the core of qualitative methods (see Chapter 4). There is a dialogue between ideas and evidence that culminates in a meaningful representation of the research subject. The main difference is that in qualitative research the emphasis is on clarifying a category and enriching its representation, whereas in comparative research the emphasis is on using contrasts between cases to further the researcher's understanding of their diversity.

Using Comparative Methods

Comparative methods are used to study configurations. A configuration is a specific combination of attributes that is common to a number of cases. For example, if all the countries that experienced mass demonstrations in response to IMF-mandated austerity were similar in having low levels of economic development, high levels of urbanization, undemocratic governments, and poorly organized opposition groups, this would constitute a specific configuration of conditions associated with mass demonstrations as a response to IMF-mandated austerity. The examination of patterns of diversity essentially involves a search for combinations of conditions that distinguish categories of cases. Thus, researchers look for uniformity within categories and contrasts between categories in combinations of conditions.

Data procedures appropriate for the study of configurations, formalized by Drass and Ragin (1989), constitute the core of the comparative approach to diversity. Comparative methods are used to examine complex patterns of similarities and differences across a range of cases. Like quantitative methods (see Chapter 6), comparative methods are used to examine causes and effects, but the emphasis in comparative research is on the analysis of configurations of causal conditions.

Before examining data procedures specific to comparative methods, first consider an example that shows the main ideas behind the techniques.

An Overview of Comparative Methods

An example from the study of the repression of austerity protests is used to illustrate general features of comparative methods. Table 5.1 presents hypothetical data on sixteen countries that experienced austerity protests in the early 1980s. Eight of these countries had governments that became violently repressive in response to austerity protests; the governments of the other eight did not.

The table shows differences and similarities among these sixteen countries with respect to conditions believed to be relevant to repression, derived from an analytic frame for government repression. The conditions include:

- whether the country was politically aligned with the Soviet Union or with the United States and Western Europe in the 1980s
- whether or not the country had undergone substantial industrialization prior to 1980
- whether or not the country had a democratic government prior to the emergence of austerity protests
- whether or not the country had a strong military establishment prior to the emergence of austerity protests

The goal of comparative analysis is to determine the combinations of causal conditions that differentiate sets of cases. In this analysis, the goal is to find combinations of casual conditions that distinguish the eight countries with governments that became repressive from the other eight countries. Careful examination of the similarities among the countries with violently repressive governments shows that they do not share any single causal condition or any single combination of conditions. However, there are two combinations of conditions that are present in the set of countries that had repressive governments that are both absent from the set that did not. The sixteen cases are sorted in the table to highlight these two combinations.

The first four cases share an absence of democratic government prior to the emergence of austerity protests combined with a strong military establishment. None of the cases in the lower half of the table (the eight countries lacking violent repression) has this combination. The second four countries with violent repression share two different conditions: a presence of democratic government prior to austerity protests combined with an absence of significant industrialization prior to the protests. Again, none of the eight countries lacking violent repression has this combination of conditions.

TABLE 5.1

Simple Example of Comparative Methods*

Case	*Aligned with USSR*	*Industrialized*	*Democratic Government*	*Strong Military*	*Violent Repression*†
1	0	0	0	1	1
2	0	1	0	1	1
3	1	0	0	1	1
4	1	1	0	1	1
5	0	0	1	0	1
6	0	0	1	1	1
7	1	0	1	0	1
8	1	0	1	1	1
9	0	0	0	0	0
10	0	1	0	0	0
11	0	1	1	0	0
12	0	1	1	1	0
13	1	0	0	0	0
14	1	1	0	0	0
15	1	1	1	0	0
16	1	1	1	1	0

*In the columns with causal or outcome conditions, the number 1 indicates the presence of a condition or "yes"; 0 indicates its absence or "no."

†The two combinations of conditions linked to violent repression are (1) absence of democratic government combined with a strong military and (2) presence of a democratic government combined with an absence of industrialization.

The results of the examination of similarities and differences thus leads to the conclusion that there are two different combinations of conditions (or causal configurations) that explain the emergence of violent repression in these cases. The first configuration (nondemocratic rule combined with a strong military) suggests a situation where the military establishment has gained the upper hand in part because of the absence of checks (democratic government) on its power. The second configuration (absence of significant industrialization combined with the presence of a democratic government prior to the emergence of violent repression) suggests a situation where a breakdown of democratic rule occurred in countries that lacked many of the social structures associated with industrialization (for example, urbanization, literacy, and so on). These social structures are believed to facilitate stable democratic rule. Further research might show important differences between these two sets of cases with respect to the kind of repression that was inflicted on the protesters.

The cases are arranged in Table 5.1 so that the main patterns of similarity among the countries with violent repression are easy to detect, and the comparison of these cases with countries lacking violent repression is simplified. Specific procedures for assessing patterns of similarity and difference are detailed in the next section. Before examining these procedures, consider several general features of the comparative analysis just presented.

1. Comparative analysis proceeds by comparing configurations of causes—*rows* of the table—and not by comparing the presence or absence of each causal condition (that is, each of the first four columns) with the presence or absence of the outcome (the last column—repression).
2. The comparative approach allows for the possibility that there may be several combinations of conditions that generate the same general outcome (government repression in the example).
3. Comparative methods can address complex and seemingly contradictory patterns of causation. One causal condition (democratic government prior to the emergence of violent repression) is important in both its present and absent condition—it appears in both causal configurations—but contributes in opposite ways.
4. The comparative approach can eliminate irrelevant causes. One causal condition (whether nor not the country was aligned with the Soviet Union) was eliminated as an important causal condition. Even though it was considered a possible factor at the outset, examination of similarities and differences among repressive and nonrepressive cases shows that this cause is not an essential part of either of the key causal combinations.

The findings in Table 5.1 are easy to see. Usually, however, the patterns are not so simple, and researchers must use more systematic comparative methods to help them analyze similarities and differences. These techniques, explained in the next sections, make it possible for researchers to find patterns that they would probably miss if they tried to unravel differences simply by "eyeballing" their cases.

Specifying Causes and Outcomes

In the comparative approach each case is understood as a combination of causal conditions linked to a particular outcome. Thus, the selection of the outcome to be studied and the specification of causal conditions relevant to that outcome are crucially important parts of a comparative investigation.

Generally, in order to specify causes, the investigator must be familiar with the research literature on the outcome (for example, "government repression") and with the cases included in the study. In this early phase of the research, the investigator explores connections between social scientific thinking (for example, about government repression) and the evidence. These early explorations lead to a clarification of the nature of the outcome to be studied and a specification of the relevant causes.

The comparative methods described in this chapter use what social scientists call **presence-absence dichotomies**. This means that causal conditions and outcomes are either present or absent in each case and can be coded "yes" or "no," as in Table 5.1. Thus, instead of using a precise measurement of industrialization (for example, the percentage of the work force employed in manufacturing) in the data analysis, an assessment might be made of whether or not substantial industrialization occurred before a specific year (again, as in Table 5.1). The use of presence-absence dichotomies simplifies the representation of cases as configurations of causes. Research methods that focus explicitly on conditions that vary by degree or level are discussed in Chapter 6.

In comparative analysis the number of causal conditions determines the number of combinations of causal conditions that are possible. For example, the specification of four causal conditions (as in Table 5.1) provides for 16 (that is, 2^4) logically possible combinations of causal conditions. Specification of five causal conditions provides for 32 (2^5) combinations, six causal conditions provides for 64 (2^6) combinations, and so on. Causal conditions are not examined separately, as in studies focusing on covariation (see Chapter 6), but in combinations.

Once causal conditions have been selected, cases conforming to each combination of causal conditions are examined to see if they agree on the outcome. In Table 5.1, there is only one case for each combination of causal conditions, so there is no possibility of disagreement. But what if there were two cases in the first row (that is, two countries that combined absence of alignment with the Soviet Union, absence of substantial industrialization before 1980, absence of democratic government, and presence of a strong military), but in one country protesters suffered violent repression while in the other they did not? The researcher would have to determine what additional factor (present in one country but absent in the other) caused repression. This new causal condition would then be added to the table for all cases.

If there are many causal combinations with cases that disagree on the outcome, then the investigator should take this as a sign that the specification of causal conditions is incorrect or incomplete. The close examination of cases that have the same presence-absence values on all the causal conditions yet have different outcomes is used as a basis for selecting additional causal variables. Investigators move back and forth between specification of causal conditions (using social science theory and their general substantive knowledge as guides) and examination of evidence to resolve these differences.

Constructing the Truth Table

Once a satisfactory set of causal conditions for a particular outcome has been identified, evidence on cases can be represented in **truth tables**. The use of truth tables facilitates the analysis of patterns of similarities and differences.

The first step in constructing a truth table is simply to list the evidence on the cases in the form of a data table. Consider for example, the data presented in Table 5.2. This table shows hypothetical evidence on thirty suburban school districts surrounding a major metropolitan area. The outcome of interest here is whether or not the elementary schools in each district track students according to ability. When students are tracked, they are grouped together into relatively homogeneous classes. Students who learn things quickly are assigned to one class, while students who learn things at an average speed are assigned to another, and so on.

Having students of uniform ability together in the same room is thought to simplify teaching, making it more efficient. After all, it clearly would be a mistake to put first graders and sixth graders in the same room. Why not apply this same principle to students within grade levels? The usual objection is that students who are assigned to the "slow" group become branded low achievers and are rarely given the opportunity to prove otherwise. Plus, being surrounded by "faster" students can motivate a "slow" student to learn faster. Assigning students to the slow group may seal their academic fate.

TABLE 5.2

Hypothetical Data on Tracking in School Districts*

School District	*Racial Diversity*	*Class Diversity*	*Competitive Elections*	*Unionized Teachers*	*Ability Tracking*
1	0	0	0	0	0
2	0	0	0	0	0
3	0	0	0	0	0
4	0	0	0	1	1
5	0	0	0	1	1
6	0	0	1	0	0
7	0	0	1	1	1
8	0	1	0	0	0
9	0	1	0	0	0
10	0	1	0	0	0
11	0	1	0	0	0
12	0	1	0	1	1
13	0	1	1	0	0
14	0	1	1	1	1
15	1	0	0	0	1
16	1	0	0	0	1
17	1	0	0	1	1
18	1	0	0	1	1
19	1	0	0	1	1
20	1	0	0	1	1
21	1	0	1	0	0
22	1	0	1	0	0
23	1	0	1	0	0
24	1	0	1	1	0
25	1	1	0	0	1
26	1	1	0	1	1
27	1	1	0	1	1
28	1	1	1	0	0
29	1	1	1	1	0
30	1	1	1	1	0

*In the columns with causal or outcome conditions, the number 1 indicates the presence of a condition or "yes"; 0 indicates its absence or "no."

The researcher in this example wanted to understand why some school districts track elementary school students and others don't. The table lists the causal conditions that the researcher, on the basis of an examination of the relevant research literatures, thought might be important:

1. whether the school district is racially diverse or predominantly white
2. whether or not the school district has a broad representation of income groups (poor, working class, middle class, and upper middle class)
3. whether or not the school board elections in the district are open and competitive, with good voter turnout
4. whether or not the teachers in the district are unionized

The first two factors (racial and class diversity) show the social composition of school districts. These factors are important because where there is more diversity, members of dominant groups (for example, whites in racially diverse districts) generally believe that tracking will benefit their children most. The competitiveness of school board elections is important because the majority of voters usually disapprove of tracking in elementary schools. They believe this practice benefits only a minority of students. In districts where school board elections are routine matters that attract little voter interest, however, the minority of families that benefit from tracking might have more influence. Unionization of teachers is included because the researcher believes that teacher unions prefer tracking because it simplifies teaching.

The school districts are sorted in Table 5.2 according to the four causal conditions, so that districts that are identical on these factors are next to each other. Inspection of the data shows that there are no districts that have the same combination of scores on the causal conditions but different outcomes. Districts 8–11, for example, all show the same pattern on the four causal conditions; they also are identical on the outcome—none of these districts tracks students according to ability. If the cases were not consistent on the outcome, it would be necessary to examine them closely to determine which other causal factors should be added to the table.

Listing the data on the cases, as shown in Table 5.2, is a necessary preliminary to the construction of the truth table. The idea behind a truth table is simple. The focus is on causal combinations. Each logical combination of values on the causal conditions is represented as one row of the truth table. Thus, truth tables have as many rows as there are logically possible combinations of values on the causal conditions. If there are four dichotomous causal conditions, as in Table 5.2, the truth table will contain $2^4 = 16$ rows. Each row of the truth table is assigned an outcome score (1 or 0, for presence–absence of the outcome) based on the cases in that row. The first three cases in Table 5.2, for example, have the same combination of scores on the causal conditions (absent on each of the four conditions) and the same outcome (absence of tracking). They are combined to form the first row of the truth table presented in Table 5.3. The number of districts that make up each row of the truth table is also reported in Table 5.3, so that the translation of Table 5.2 to Table 5.3 is clear.

Simplifying the Truth Table

The truth table (Table 5.3) summarizes the causal configurations that exist in a data table (Table 5.2). Listing configurations is not the same as identifying patterns, however. Usually, comparative researchers want to examine configurations to see if they can be simplified. When investigators simplify configurations, they identify patterns.

A quick example of simplification: Look at rows 13 and 14 of the truth table reported in Table 5.3. Row 13 reports that school districts that combine the following four characteristics track students: (1) racial diversity, (2) class diversity, (3) an absence of competitive school board elections, and (4) an absence of teachers' unions. Row 14 reports that school districts that differed on only one of these four conditions—teachers' unions—also tracked students. The comparison of these two rows shows that when the first two causal conditions are present (race and class diversity) and the third is absent (competitive school board elections), it does not matter whether teachers are unionized; tracking by ability still takes place.

An easy way to represent this simplification is to use uppercase letters to indicate presence of a condition and lowercase letters to indicate its absence. In this example, the word "RACE" indicates the presence of racial diversity; "race" is used to indicate its absence. "CLASS" is used to indicate the presence of class diversity, "class" to indicate its absence. "ELECTIONS" is used to indicate the presence of open, competitive school board elections, "elections" to indicate the absence of this condition. "UNIONS" indicates the presence of teachers' unions, "unions" the absence of this condition. Finally, "TRACKING" indicates the presence of tracking, and "tracking" its absence.

TABLE 5.3

Truth Table for Data on Tracking in School Districts*

Row	*Racial Diversity*	*Class Diversity*	*Competitive Elections*	*Unionized Teachers*	*Ability Tracking*	*Number of Districts*†
1	0	0	0	0	0	3
2	0	0	0	1	1	2
3	0	0	1	0	0	1
4	0	0	1	1	1	1
5	0	1	0	0	0	4
6	0	1	0	1	1	1
7	0	1	1	0	0	1
8	0	1	1	1	1	1
9	1	0	0	0	1	2
10	1	0	0	1	1	4
11	1	0	1	0	0	3
12	1	0	1	1	0	1
13	1	1	0	0	1	1
14	1	1	0	1	1	2
15	1	1	1	0	0	1
16	1	1	1	1	0	2

*In the columns with causal or outcome conditions, the number 1 indicates the presence of a condition or "yes"; 0 indicates its absence or "no."

†The number of districts is reported simply to remind the reader that each row of a truth table may represent more than one case.

Thus, row 13 can be represented as

TRACKING = RACE·CLASS·elections·unions

and row 14 as:

TRACKING = RACE·CLASS·elections·UNIONS

where multiplication (·) is used to indicate the combination of conditions. These two rows can be simplified through combination because they have the same outcome and differ on only one causal condition—presence–absence of teachers' unions. This simplification strategy follows the logic of an experiment. Only one condition at a time is allowed to vary (the "experimental" condition). If varying this condition has no discernible impact on the outcome, it can be eliminated as a factor. Thus, the comparison of rows 13 and 14 results in a simpler combination:

TRACKING = RACE·CLASS·elections

This rule for combining rows of the truth table as a way of simplifying them can be stated formally: If two rows of a truth table differ on only one causal condition yet result in the same outcome, then the causal condition that distinguishes the two rows can be considered irrelevant and can be removed to create a simpler combination of casual conditions (a simpler term).

The process of combining rows to create simpler terms can be carried on until no more simplification is possible. Table 5.4 shows all the simplifications that are possible for the truth table in Table 5.3, using presence of ability tracking as the outcome of interest. In Table 5.4 the truth table rows from Table 5.3 with outcomes of "1" (presence of tracking) have been translated into the uppercase and lowercase names in the manner just described. Panel A of this table simply lists the eight kinds of districts that track students according to ability. Panel B shows the first round of simplification. Each of the terms from panel A can be combined with one or more other terms to create simpler terms. Whenever two terms with four conditions are combined, the new term has three conditions because one condition has been eliminated.

Panel C shows the second round of simplification. In this round, terms with three conditions (from panel B) are combined to form terms with two conditions. For example, the term labeled #17 in panel B (race·class·UNIONS) can be combined with the term #21 (race·CLASS·UNIONS) to form a two-condition term (race·UNIONS). All the terms from panel B combine with one or more terms from the same panel to produce the three two-condition terms listed in panel C.

The three terms in panel C can be represented in a single statement describing the conditions under which tracking in these suburban school districts occurs:

TRACKING = race·UNIONS + RACE·elections + elections·UNIONS

Tracking occurs when:

1. racial diversity is absent and teachers' unions are present,
2. racial diversity is present and competitive school board elections are absent, or
3. competitive school board elections are absent and teachers' unions are present.

TABLE 5.4

Simplification of Truth Table for Tracking (Table 5.3)

Panel A. Districts That Track Students

Row	*Causal Configurations*
2	race·class·elections·UNIONS
4	race·class·ELECTIONS·UNIONS
6	race·CLASS·elections·UNIONS
8	race·CLASS·ELECTIONS·UNIONS
9	RACE·class·elections·unions
10	RACE·class·elections·UNIONS
13	RACE·CLASS·elections·unions
14	RACE·CLASS·elections·UNIONS

Panel B. First Round of Simplification

			Label for New Term
Rows 2 + 4	→	race·class·UNIONS	#17
Rows 2 + 6	→	race·elections·UNIONS	#18
Rows 2 + 10	→	class·elections·UNIONS	#19
Rows 4 + 8	→	race·ELECTIONS·UNIONS	#20
Rows 6 + 8	→	race·CLASS·UNIONS	#21
Rows 6 + 14	→	CLASS·elections·UNIONS	#22
Rows 9 + 10	→	RACE·class·elections	#23
Rows 9 + 13	→	RACE·elections·unions	#24
Rows 10 + 14	→	RACE·elections·UNIONS	#25
Rows 13 + 14	→	RACE·CLASS·elections	#26

Panel C. Second Round of Simplification

#17 + #21	→	race·UNIONS
#18 + #20	→	race·UNIONS
#18 + #25	→	elections·UNIONS
#19 + #22	→	elections·UNIONS
#23 + #26	→	RACE·elections
#24 + #25	→	RACE·elections

Before accepting these tentative results, it is important to determine if further simplification is possible, as is often the case. Sometimes the process of combining rows to produce simpler terms (presented in Table 5.4) generates "surplus" terms. A surplus term is redundant with other terms and is not needed in the statement describing the combinations of conditions linked to an outcome. In short, some of the terms that are left after the process of combining rows may be superfluous. Recall that the goal of comparative analysis is to describe diversity in a simple way. If the results can be further simplified by eliminating surplus terms, as is the case here, it is important to do so. The idea of a surplus term is best understood by examining the methods used to detect them.

The best way to check if there are surplus terms is to construct a chart showing which of the original terms in panel A are covered by which simplified terms in panel C. A simplified term covers a truth table row if the row is a subset of the simplified term. For example, RACE·CLASS·elections·UNIONS (row 14 of the truth table) is a subset of the simplified term elections·UNIONS.

The chart showing the coverage of the simplified terms is presented in Table 5.5. The simplified term race·UNIONS covers the first four terms from panel A of Table 5.4, while the term RACE·elections covers the other four. The third simplified term (elections·UNIONS) does not cover any of the rows uniquely; it covers two that are covered by the first simplified term and two that are covered by the second. Thus, the third simplified term is surplus; it is redundant with the other terms.

By eliminating the third simplified term the results of the analysis of configurations can be reduced to

TRACKING = race·UNIONS + RACE· elections

This completes the procedure. The final statement says that tracking occurs (1) when racial diversity is absent and teachers' unions are present, or (2) when racial diversity is present and competitive school board elections are absent. The first term indicates that in school districts that are predominantly white, tracking is implemented if there are teachers' unions. This finding supports the researcher's belief that teachers' unions prefer tracking and specifies the conditions under which their interests are realized—in districts where there is an absence of racial diversity. It does not matter whether school board elections are open and competitive or whether the district contains a broad range of income groups. The second term indicates that in school districts where there is racial diversity, tracking occurs when school board elections are not competitive. They are routine matters that do not attract a lot of voter interest. In these districts, it does not matter whether teachers' unions are present or whether the districts contain a broad range of income groups. The second term suggests that if voters become involved in school board elections, tracking would be eliminated in racially diverse districts.

TABLE 5.5

Chart Showing Coverage of Simplified Terms

	Simplified Terms[†]		
*Truth Table Rows**	*race·UNIONS*	*RACE·elections*	*elections·UNIONS*
race·class·elections·UNIONS	X		X
race·class·ELECTIONS·UNIONS	X		
race·CLASS·elections·UNIONS	X		X
race·CLASS·ELECTIONS·UNIONS	X		
RACE·class·elections·unions		X	
RACE·class·elections·UNIONS		X	X
RACE·CLASS·elections·unions		X	
RACE·CLASS·elections·UNIONS		X	X

*From panel A of Table 5.4.

[†]From panel C of Table 5.4.

The analysis of school districts presented here shows the major steps in using comparative techniques to unravel causal patterns.

1. Select causal and outcome conditions, using existing social science literature and substantive knowledge to guide the selection.
2. Construct a sorted data table showing the scores of cases on these causal and outcome conditions (Table 5.2).
3. Construct a truth table from the data table, making sure that cases with the same causal conditions actually have the same score on the outcome (Table 5.3).
4. Compare rows of the truth table and simplify them, eliminating one condition at a time from pairs of rows (Table 5.4).
5. Examine the coverage of the simplified terms to see if there are any surplus terms that can be eliminated (Table 5.5).

The terms that remain after step 5 show the simplest way to represent the patterns of diversity in the data. In the comparative analysis presented in Tables 5.2 through 5.5, the goal is to explain why some school districts track elementary students. The results show which types of school districts track elementary students and distinguishes them from those that do not.

Conclusion

The brief overview of comparative methods presented in this chapter illustrates some of the key features of the comparative approach. The most important feature is its focus on diversity. Whenever a set of cases have different outcomes (cities with different reactions to Indochinese refugees, countries with different reactions to austerity programs, bars with different ways of encouraging patrons to drink or to not drink, and so on), comparative methods can be used to find simple ways of representing the patterns of diversity that exist among the cases. These methods identify similarities within subsets of cases that distinguish them from other subsets.

As in all forms of social research, analytic frames and images play an important part in comparative research. Analytic frames provide primary leads for the construction of truth tables, especially the selection of causal conditions. The construction of the truth table itself is an important part of the dialogue of ideas and evidence in comparative research because the truth table must be free of inconsistencies before it can be simplified. Evidence-based images emerge from the simplification of truth tables in the form of configurations of conditions that differentiate subsets of cases.

In many ways the comparative approach lies halfway between the qualitative approach and the quantitative approach. The qualitative approach seeks in-depth knowledge of a relatively small number of cases. When the focus is on commonalities, it often narrows its scope to smaller sets of cases as it seeks to clarify their similarities. The comparative approach usually addresses more cases because of its emphasis on diversity, and it is applied to sets of cases that are clearly bounded in time and space. As Chapter 6 shows, the quantitative study of covariation seeks broad familiarity with a large number of cases and most often views them as generic, interchangeable observations.

6

Using Quantitative Methods to Study Covariation

Introduction

The starting point of quantitative analysis is the idea that the best route to understanding basic patterns and relationships is to examine phenomena across many cases. Focusing on any single case or on a small number of cases might give a very distorted picture. Looking across many cases makes it possible to average out the peculiarities of individual cases and to construct a picture of social life that is purified of phenomena that are specific to any case or to a small group of cases. Only the general pattern remains.

Quantitative researchers construct images by showing the covariation between two or more features or attributes (variables) across many cases. Suppose a researcher were to demonstrate in a study of the top 500 corporations that those offering better retirement benefits tend to pay lower wages. The image that emerges is that corporations make trade-offs between retirement benefits and pay, with some corporations investing in long-term commitments to workers (retirement benefits) and some emphasizing short-term payoffs (wages and salaries). Evidence-based images such as these are general because they describe patterns across many cases and they are *parsimonious*—only a few attributes or variables are involved (pay and retirement benefits).

Images that are constructed from broad patterns of covariation are considered general because they condense evidence on many cases. The greater the number of cases, the more general the pattern. A quantitative researcher might construct a general image of political radicalism that links degree of radicalism to some other individual-level attribute, such as degree of insulation from popular culture, and use survey data on thousands of people (including people who are politically inert) to document the connection. Qualitative researchers studying this same question would go about the task very differently. The images they construct are detailed and specific, and they use methods that enhance rather than condense evidence. Using a qualitative approach, a researcher might construct an image of how political radicals nurture their radical commitments by studying the daily lives of twenty radicals in depth.

These two images of radicalism, one by a qualitative researcher and one by a quantitative researcher, might or might not contradict. Even if they did not contradict each other, the two images still would be very different in degree of detail and complexity. Quantitative researchers sacrifice in-depth knowledge of each individual case in order to achieve an understanding of broad patterns of covariation across many cases.

Quantitative researchers often use the term *correlation* to describe a pattern of covariation between two measurable variables. In the previous example, degree of radicalism and degree of insulation from popular culture are correlated such that more radical people tend to be more insulated. They also sometimes describe a correlation between two variables as a relationship, which should not be confused with the more conventional use of the term *relationship* to describe social bonds (for example, two lovers have a relationship). Again using the previous example, there is a relationship between degree of radicalism and degree of insulation.

Usually, attributes of cases that can be linked in this way are understood as variables because they are phenomena that vary by level or degree. There are cases with high values of a variable (for example, more than eighteen years of education on the variable "educational attainment"), cases with moderate values (say, twelve years of education), and cases with low values (only a few years of education). Some variables (called **independent** or causal variables) may be defined as causes, and others (called **dependent** or outcome variables) may be defined as effects in a given analysis. The dependent variable is the phenomenon the investigator wishes to explain; independent variables are the factors that are used to account for the variation in the dependent variable. A dependent variable in one analysis (for example, Gross National Product per capita in a study that seeks to explain why some countries are poor and others rich) may appear as an independent variable in the next (for example, as a causal variable that explains why people in some countries have a higher life expectancy than people in other countries).

The Goals of Quantitative Research

Because the quantitative approach favors general features across many cases, it is especially well suited for several of the basic goals of social research. These include the goals of identifying general patterns and retionships, testing theories, and making predictions. These three goals all dictate examination of many cases—the more, the better—and favor a dialogue of ideas and evidence that centers on how attributes of cases (variables) are linked to each other.

Identifying General Patterns and Relationships

One of the primary goals of social research is to identify general relationships. For a relationship to be general, it must be observed across many cases. In quantitative research this is understood not as observing the same exact phenomenon in each and every case, but as observing an association between two or more phenomena across many cases. When a social researcher claims that poorer countries tend to have higher rates of homicide, he or she in essence is stating that there is a general correspondence between a country's wealth and its rate of homicide such that richer countries tend to have lower homicide rates and poorer countries tend to have higher rates. (The United States is a striking exception to this general relationship.)

Identifying general patterns and relationships is important because they offer important clues about causation. It is obviously not true that if two variables are related across many cases, then one necessarily causes the other. If we found that shoe size and income were related, we would not argue that big feet cause high incomes. However, when variables are systematically related, it is important to consider the *possibility* that one may cause the other. Alternatively, the two correlated variables both may be the effects of some third, unidentified variable.

An example: In the United States over most of the twentieth century, the more industrial states have tended to offer stronger support for liberal Democratic candidates. This general pattern connects an independent variable, percentage of the state's adult population employed in industry, to a dependent variable, percentage of a state's electorate voting for liberal Democratic candidates. A causal relationship can be inferred from the correlation between these two variables: Conditions associated with having a lot of industry (such as urbanization, unionization, and so on) generate a preference for the liberal candidates among the people affected by these conditions. The explanation of liberal voting based on this evidence thus may emphasize the impact of industrial conditions on people's interests and the translation of these interests to a preference for liberal candidates. The causal images behind correlations are central to the representations of social life that quantitative researchers construct.

Generally, quantitative social researchers identify causation with explanation. Once the causes of a phenomenon have been identified, it has been explained. The usual sequence is:

1. a pattern of covariation is identified and the strength of the correlation is assessed,
2. causation may be inferred from the correlation, and, if so,
3. an explanation is built up from the inferred causal relationship.

Another way of understanding this is simply to say that quantitative social researchers construct images by examining patterns of covariation among variables and inferring causation from these broad patterns.

Testing Theories

While quantitative researchers often construct explanations and images from the broad patterns that they observe (like the rough correlation between income levels and educational levels) and relate these evidence-based images to their ideas about social life, they also test ideas drawn directly from social theories. Recall from Part I of this book that all social researchers are involved in long-standing, abstract conversations about social life. Social researchers use this body of thought whenever they construct images, but they also seek to advance this body of thought and to construct formal tests of ideas drawn from it.

Testing an idea is different from *using* an idea to help make sense of some pattern in a set of data or body of evidence that already has been collected. When an idea is tested, it is first used to construct an image that is based on the ideas themselves, not the evidence. The researcher constructs a theoretical image. Researchers use these theoretically based images to derive testable propositions (also called hypotheses) about evidence that has not yet been examined. Once examined, the evidence either supports or refutes the proposition (see Chapter 1).

This formal assessment of hypotheses helps social scientists determine which ideas are most useful for understanding social life. An idea that consistently fails to win support in these formal tests will eventually be dropped from the pool of ideas that social scientists use. Ideas that consistently receive support are retained.

One theoretical image in the study of social inequality is the idea that advanced societies are *achievement* oriented—they reward performance, while less advanced societies are *ascription* oriented—they reward people for who they are (for example, their family's social status). Thus, in an achievement-oriented society, a person of great ability from a low-status, impoverished background should nevertheless be successful. By contrast, in an ascription-oriented society, people born into high-status families will be successful, regardless of their talents.

These are theoretical images. There is no society that is totally achievement oriented, nor is there any society that is totally ascription oriented. However, these theoretical images have implications for inequality in the United States, which is generally considered to be an advanced society (despite its absurdly high homicide rate). Has the United States become more achievement oriented over the last forty years? Is it easier today for a talented person from a low-status, impoverished background to succeed than it was in the 1950s? The theoretical images just described link the ascendance of the achievement orientation to societal advancement, suggesting that over the last forty years it should have become easier in the United States for a talented person from a low-status background to get ahead.

Thus, the testable proposition is that evidence on "social mobility" (the study of who gets ahead) should support the idea that achievement has become more important and ascription less important in U.S. society. The increased importance of achievement criteria might be discernible in the strength of the relationship between educational achievement and subsequent income. Is the correlation between these two variables stronger in 1994 than it was in 1954? The decreased importance of ascription might be visible in the strength of the relationship between race and income. Is being black less of a liability in 1994 than it was in 1954? Of course, it would be possible to examine the effects of a variety of achievement and ascription variables on income over the last forty years (and at various points within this span of time) because there have been many surveys conducted over this period with data relevant to the proposition.

The quantitative approach is very useful for testing theoretical ideas and images such as these. Notice that these ideas are *general*—they are relevant to many cases, and they are *parsimonious*—they concern the operation of only a few causal variables. When theoretical ideas are relevant to many cases, like ideas about ascription versus achievement, we have more confidence in a test when it includes a very large number and a wide range of cases.

Making Predictions

Another goal of social research that mandates examination of large numbers of cases is making predictions. In order to be able to make predictions it is important to have as many cases as possible and to have a variety of cases. When predictions are based on many cases, researchers have the largest possible data base at their disposal and are capable of making the most accurate predictions.

For example, to predict whether middle-aged, middle class, white, Southern males will favor the Republican candidate in the next presidential election, it is necessary to know how people with this combination of characteristics generally vote in presidential elections. Do they always favor Republican candidates? Do they vote differently when the Democratic candidate is a Southerner? When issues related to national defense are important, are they more enthusiastic in their support for the Republican candidate? Clearly, the greater the volume of evidence on the political behavior of males in this category, the more precise the prediction for a future election.

Having a lot of evidence makes it easier to forecast future behavior. Knowledge of general patterns also helps. Suppose a researcher wants to predict the political behavior of middle-aged, middle class, Southern white males in an election that pits a Democratic candidate from the South against a Republican candidate who favors greater military spending. Suppose further that this particular combination of candidate characteristics has never occurred before. How can social scientists extrapolate when one condition (Democratic candidate from the South) decreases this group's support for the Republican candidate, while the other (a pro-military posture) increases its support?

Accumulated knowledge of general patterns helps in these situations. If research shows that, in general, the personal characteristics of a candidate (for example, being a Southerner) matter more to voters than the positions a candidate takes (for example, being pro-military), then the prediction would be that the Southern factor should outweigh the military factor.

Knowledge of general patterns helps social researchers sharpen their predictions by providing important clues about how to weight factors accurately, even in the face of many unknowns and great uncertainty. Because it is well suited for the production and accumulation of knowledge about general patterns, the variable-based approach offers a solid basis for making such predictions.

Contrasts with Qualitative and Comparative Research

When social researchers construct images from evidence, they may use any number of cases. Qualitative researchers typically use a small number of cases (from one to several handfuls); comparative researchers use a moderate number; and quantitative researchers use many (sometimes thousands). The images that qualitative researchers construct are detailed and in depth; the images that quantitative researchers construct are based on general patterns of variation across many, many cases. These general images link variation in one attribute of cases to variation in other attributes. The patterns of covariation between two or more such variables across many cases provide the basic raw material for the images that quantitative researchers construct.

The quantitative strategy favors **generality.** A quantitative researcher might show that there is a link between variation in income levels and variation in educational levels in a large sample of U.S. adults. This pattern of covariation evokes a general image of how people in the United States get ahead. If income levels covary more closely with educational levels than they do with other individual-level attributes (such as age, race, marital status, and so on), then it appears that success in the educational system is the key to subsequent material well-being. This image of how income differences arise in U.S. society is very different from one that links differences in income levels to differences in other attributes such as skin color. A key question in the application of the quantitative approach is the strength of the correlation of different causal variables, like educational level and skin color, to dependent variables, like income.

The quantitative approach prizes not only generality, but also **parsimony**—using as few variables as possible to explain as much as possible. In a study of income levels, for example, the main concern of the quantitative researcher would be to identify the individual-level attributes with the strongest correlation with income levels. Is it educational levels? Is it age? Is it parents' income? Is it skin color? Which variables have the strongest links with differences in income? By identifying the variables with the strongest correlations, quantitative researchers pinpoint key causal factors and use these to construct parsimonious images.

Parsimony and generality go together in quantitative research. Images that are general also tend to be parsimonious. It is clear that parsimony is not a key concern of the qualitative approach. Qualitative researchers believe that in order to represent subjects properly, they must be studied in depth—to uncover nuances and subtleties. Comparative researchers lie halfway in between on the issues of parsimony and generality. Rather than focus on patterns that are general across as many cases as possible—the primary concern of the quantitative approach, comparative researchers focus on diversity, on configurations of similarities and differences within a specific set of cases.

This difference between quantitative and comparative research is subtle but important. A parsimonious image that links attributes across many cases assumes that all cases are more or less the same in how they came to be the way they are. The person with low education and low income is, in this view, the reverse image of the person with high education and high income. They are two sides of a single coin.

The comparative approach, by contrast, focuses on diversity—how different causes combine in complex and sometimes contradictory ways to produce different outcomes. Thus, instead of focusing on attributes that covary with differences in income levels, like educational levels, the comparative researcher might focus on the diverse ways people achieve material success, with and without education, and contrast these with the diverse ways they fail to achieve success. From a comparative perspective, it is not a question of which attributes covary most closely with income levels, but of the different paths to achieving material success.

Of course, the comparative approach is best suited for the study of a moderate number of cases, not for the study of income differences across thousands of cases. Like the qualitative approach, the comparative approach values knowledge of individual cases. The important point in this contrast between the quantitative approach and the comparative approach is the difference between looking for variables that seem to be systematically linked to each other across many cases (a central concern of the quantitative approach) and examining patterns of diversity (a major objective of the comparative approach).

The Process of Quantitative Research

The quantitative approach is the most structured of the three research strategies examined in this book. Its structured nature follows in part from the fact that it is well suited for testing theories. Whenever researchers test theories, they must exercise a great deal of caution in how they conduct their tests so that they do not rig their results in advance. Human beings are reactive creatures. There is a large body of research showing that when people are interviewed, their responses are shaped in part by the personal characteristics of the interviewer (such as whether the interviewer is male or female). If they know what a social scientist is trying to prove, they may try to undermine the study, or they may become overcompliant. Tests in any scientific field that are not conducted carefully cannot be trusted.

The more structured nature of quantitative research also follows from its emphasis on variables. Variables are the building blocks of the images that quantitative researchers construct. But before researchers have variables that they can connect through correlations, they must be able to specify their cases as members of a meaningful set, and they must be able to specify the aspects of their cases that are relevant to examine as variables. In short, much about the research tends to be fixed at the outset of the quantitative investigation.

This orientation contrasts sharply with those of the other two strategies. In qualitative research, investigators often do not decide what their case is a "case of" until they write up their results for publication (see Chapter 4). In the comparative approach, researchers assume that their cases are very diverse in how they came to be the way they are, and investigators often conclude their research by differentiating distinct types of cases (see Chapter 5). Of course, quantitative researchers are quite capable of differentiating types of cases, but their primary focus is on relating variables across all the cases they have data on.

Cases and variables can be fixed at the outset of a study—as they tend to be in quantitative research—only if the study is well grounded in an analytic frame. Thus, analytic frames play a very important part in quantitative research.

Analytic Frames in Quantitative Research

Researchers use analytic frames to articulate theoretical ideas about social life (see Chapter 3). Frames specify the cases relevant to a theory and delineate their major features. The importance of frames to quantitative research can be seen most clearly in research that seeks to test theories. Once a theory has been translated into an analytic frame, specific propositions (or testable hypotheses) about how variables are thought to be related to each other can be stated. Researchers can then develop measures of the relevant variables, collect data, and use correlational techniques to assess the links among relevant variables. Relationships among variables either refute or support theoretically based images.

A theory of job satisfaction may emphasize the match between a person's skills and talents, on the one hand, and the nature of the tasks he or she is required to perform, on the other. The basic theoretical idea is that people are happiest in their work when their job requires them to do things they are good at. Work that does not suit an employee makes the employee feel frustrated and dissatisfied, even useless. These theoretical ideas can be expressed in a frame that details employee and job characteristics relevant to job satisfaction.

To test the idea that job satisfaction is greatest when skills and duties are well matched, it would be necessary to elaborate this frame in advance of data collection. Of course, researchers should not remain ignorant of their research subjects before testing a theory. They should learn all that they can. The point is simply that the data used to test a theory is not the same as the evidence the researcher uses in developing or refining the hypothesis to be tested. To do this would be to rig the results of the test in a way that would confirm the researcher's ideas.

The frame becomes more or less fixed once theory testing is initiated. The job satisfaction frame is fixed on employees as cases, job satisfaction as the dependent variable, and the match between employee and job characteristics as independent variables. When a frame is fixed, the images that can be constructed from evidence are constrained. When the goal is to test theory, the images that can be constructed are further constrained by the hypothesis. In the job satisfaction example, if the researcher finds that the employees who are well matched in terms of skills and duties are not the ones with the highest levels of job satisfaction, then the image constructed from the evidence rejects the theoretically based frame.

Even when quantitative researchers are not testing theories, the images that they can construct from evidence are still constrained by their frames. In order to examine relationships among variables, it is necessary first to define relevant cases and variables. The examination of relationships among variables usually cannot begin until after all the evidence has been collected. Furthermore, the evidence that is collected must be in a form appropriate for quantitative analysis. There must be many cases, all more or less comparable to each other, and they must have data on all, or at least most, of the relevant variables. Thus, quantitative research implements frames directly, as guides to data collection, telling researchers which variables to measure.

From Analytic Frame to Data Matrix

In quantitative research the collection of evidence is seen as a process of filling in the data table (or **data matrix**) defined by the analytic frame. (An example of a small data matrix is presented in Table 6.1.) In the study of job satisfaction, the data on a single employee would fill one row of the data matrix, and there would be as many rows as employees. The columns of the data matrix would be the different employee and job characteristics relevant to the analysis. Thus, in quantitative research the data matrix mirrors the analytic frame.

The researcher would not fill in this matrix with data on just anyone. In a study of job satisfaction, for example, the researcher would probably want to collect data on all the employees of a particular factory or firm. (Of course, if the firm or factory were very large, the researcher would probably collect a systematic, random sample of its employees.) In order to construct a good test of the theory, the researcher would choose a work setting with many different kinds of jobs and with employees possessing many different kinds of skills. This combination would provide a good setting for testing the idea that matching skills with duties is important for job satisfaction. If the researcher chose a work setting where everyone did more or less the same thing and had more or less the same skills, then it would not be an appropriate setting for testing the idea that matching skills with duties matters.

Thus, quantitative researchers exercise considerable care when selecting the cases to be used for testing a particular theory. The cases must be relevant to the theory, and they must vary in ways that allow the theory to be tested. When a theory is relevant to very large numbers (for example, all adults in the United States), the quantitative researcher uses a random sample of such cases (for example, every 10,000th person listed in the census). When it is not possible to use a national sample, the researcher may sample the people in a single city or region that is representative of the population as a whole.

Of course, not all social theories are about variation among individuals. Sometimes they are about other basic units—firms, families, factories, organizations, gangs, neighborhoods, cities, households, bureaucracies, even whole countries. In most quantitative research, cases are common, generic units like these. This preference for generic units follows from its emphasis on constructing broad, parsimonious images that reflect general patterns.

Measuring Variables

Quantitative researchers also exercise great care in developing measures of their variables. In the study of job satisfaction, the measurement of the dependent variable is critically important to the study as a whole. How should it be measured? Is it enough simply to ask employees to rate their degree of satisfaction with their jobs? Can employees be trusted to give honest and accurate assessments or will they worry that management is looking over their shoulders? Should the researcher also examine personnel files? Is this legal? Is it ethical? What about records on absenteeism? Is absenteeism a good measure of job dissatisfaction? What about asking supervisors to give their ratings of the people who work under them?

Not surprisingly, there is an immense literature on the problems of measuring job satisfaction, and comparably large literatures exist on the measurement of most of the many variables that interest social scientists.

Even variables that seem straightforward are difficult to measure with precision, and controversies abound. What does years of education measure? Knowledge? Job-relevant skills? Time spent in classrooms?

For example, it is clear that nations differ in wealth. Gross National Product in U.S. dollars per capita (GNP per capita) is a conventional measure of national wealth. However, GNP per capita has important liabilities. Some are technical. In order to get all countries on the same yardstick, their currencies must be converted to U.S. dollars. But the relevant exchange rates for making these conversions fluctuate daily. Thus, the rankings of countries on GNP per capita fluctuate daily. But wealth differences between countries are thought to be relatively long standing; differences induced by short-term exchange rate fluctuations are artificial.

A more serious problem: Some countries have a high GNP per capita but do not seem wealthy because most of their citizens do not live well. In the mid-1970s, for example, the GNP per capita of many oil-exporting countries skyrocketed, but living conditions in these countries were not as good as those of some poorer, non–oil-exporting countries. Thus, it is possible, at least in the short run of a decade or so, to have a high GNP per capita and relatively poor living conditions, which contradicts the idea of GNP per capita as a measure of national wealth.

A still more serious problem: Some countries have great income inequality, with a substantial class of very rich people, many poor people, and few in between. These countries may appear to be much better off than they are because on the average—which is what GNP per capita captures—conditions seem OK. But the reality may be one of widespread suffering in the face of extreme riches.

The issue of using appropriate measures is known as the problem of validity (see also Chapter 1). Do data collection and measurement procedures work the way social researchers claim? One way to assess validity is to check the correlations among alternative measures that, according to the ideas that motivate the study, should covary. For example, a researcher may believe that years of education is a valid measure of general knowledge and could assess this by administering a test of general knowledge to a large group of people representative of the population to be surveyed. If their scores on this test correlate strongly with their years of education, then the researcher would be justified in treating years of education in the survey of the larger population as a measure of general knowledge.

Researchers are also concerned about the reliability of their measures. **Reliability** generally concerns how much randomness there is in a particular measure (quantitative researchers refer to this as *random error*). For example, day-to-day exchange rate fluctuations produce randomness in the GNP per capita in U.S. dollars. The calculation of GNP per capita in U.S. dollars changes every time exchange rates change. Thus, GNP per capita calculated one day will not correlate perfectly with GNP per capita calculated the next, even though the estimates of the goods and services produced by each country are unchanged.

Consider an example closer to home: When employees are asked how satisfied they are with their jobs, their answers may reflect what happened that day or over the last few days. Ask them again in a month, and their answers may reflect what's happening then. Thus, when the measurements of job satisfaction taken one month apart are correlated, asking the same people the same question, the relationship may be weak because of the randomness induced by different surrounding events.

Researchers have developed a variety of ways to counteract unreliability. In research on job satisfaction, they might ask many questions that get at many different aspects of job satisfaction and use these together to develop a broad measure (for example, by adding the responses to form a total score for each person). More than likely, employees' responses to many of the questions will not change over one month. Thus, by adding together the responses to many related questions on job satisfaction, the researcher might develop a measure that is more reliable.

Measurement is one of the most difficult and most important tasks facing the quantitative researcher because so much depends on accurate measurement. If a correlation is weak, say between job satisfaction and a measure of the match between employees' skills and duties, is it because the theory is wrong or because the measures are bad? Is the measure of job satisfaction accurate? Is the measure of skills adequate? Is the measure of the match of employees' skills and duties properly conceived and executed? In the quantitative approach, there is no way to know for sure why a correlation that is expected to be strong comes out weak. Because researchers usually hold fast to their theories, they often blame their measures and complain about the difficulty of measuring social phenomena with precision.

Examining Correlations and Testing Theories

The examination of correlations among variables is the core of the quantitative approach, but quantitative researchers must travel a great distance before they can compute a single correlation. They must translate their theoretical ideas into analytic frames. They must choose appropriate cases. If there are many, many such cases, they must devise a sampling strategy. They must develop valid, reliable measures of all their variables. If the goal of the investigation is to test theory, they must also articulate the proposition to be tested and take great care in measuring the variables central to the proposition. And they must fill in the data matrix defined by their analytic frames, the cases they have selected, and the measures they have devised.

After all this preparation, the computation of correlations may seem anticlimactic. In qualitative research, the investigator engages ideas in every stage of the research, refining and clarifying categories and concepts as new evidence is gathered (see Chapter 4). In comparative research, a similar process of linking ideas and evidence occurs in the construction of truth tables (see Chapter 5). In quantitative research investigators must know a lot in advance of data collection. They must learn as much as they can about the theories they want to test, about their cases, and about how to measure their variables before they collect the data that will be used to test their theories. Thus, the examination of relationships among variables (the technique quantitative researchers use to construct evidence-based images) is near the end of a very long journey.

When quantitative researchers test theories, the key question is whether or not the correlations follow patterns consistent with the ideas that motivated the study. Sometimes this assessment involves the correlation between a single independent variable and a single dependent variable. In the study of job satisfaction: How strong is the correlation between job satisfaction and the degree to which employees' skills and duties are matched? Sometimes testing a theory involves comparing the strength of a correlation in different times or settings: Is educational level more strongly linked to income level in 1994 than it was in 1954? Sometimes testing involves comparing the correlations of several independent variables with one or more dependent variables: Is the effect of race on income stronger or weaker than the effect of education on income? Did the pattern change between 1954 and 1994?

What do researchers do when correlations do not support their theories? Sometimes, they simply report that the evidence does not support their theory. In other words, they report that they attempted to construct an evidence-based image consistent with some theory, but were unable to do so, suggesting that the theory is wrong. In general, however, the audiences for social science expect social life to be represented in some way in a research report. They do not expect a report of a failed attempt to construct a representation. Such reports should be more common than they are because the logic of theory testing (that is, the effort to figure out which ideas are best supported by evidence) indicates that negative findings (that is, failed representations) are very important.

More often, if the initial test of a hypothesis fails, researchers examine their evidence closely to see if there is support for their theory under specific conditions. After finding a weak correlation between job satisfaction and the degree to which employees' skills and duties are matched, a researcher might consider the possibility that other factors need to be considered. Perhaps employees who have been with the firm the longest are more satisfied, regardless of how well their skills are matched to their duties. This factor would need to be taken into account when examining the relation between job satisfaction and the match of skills and duties. Generally, researchers try to use their general knowledge of their cases and their theoretical understanding to anticipate refinements like these *before* they collect their data. They may also specify additional hypotheses in advance as a way to anticipate such failures.

Using Quantitative Methods

An Introduction to Quantitative Methods

Quantitative methods focus directly on relationships among variables, especially the effects of causal or *independent* variables on outcome or *dependent* variables. Another way to think about the quantitative approach is to see the level of the dependent variable (for example, variation across countries in life expectancy) as something that *depends on* the level of other variables (for example, variation across countries in nutrition). The strength of the correlation between the independent and the dependent variable provides evidence in favor of or against the idea that two variables are causally connected or linked in some other way.

The exact degree to which two variables correlate can be determined by computing a **correlation coefficient**. The most common correlation coefficient is known as Pearson's r and is the main focus of this discussion. If the correlation is substantial and the implied cause–effect sequence makes sense, then the cause (the independent variable) is said to "explain variation" in the effect (the dependent variable).

If cities in the United States with lower unemployment rates also tend to have lower crime rates, then these two features of cities, unemployment rates and crime rates, go together; they correlate. Generally, social scientists would argue that the unemployment rate (the independent variable) explains variation across cities in the crime rate (the dependent variable). The general pattern of covariation in this hypothetical example is high unemployment rates–high crime rates, moderate unemployment rates–moderate crime rates, and low unemployment rates–low crime

FIGURE 6.1

Plot of Crime Rate with Rate of Unemployment Showing Positive Correlation

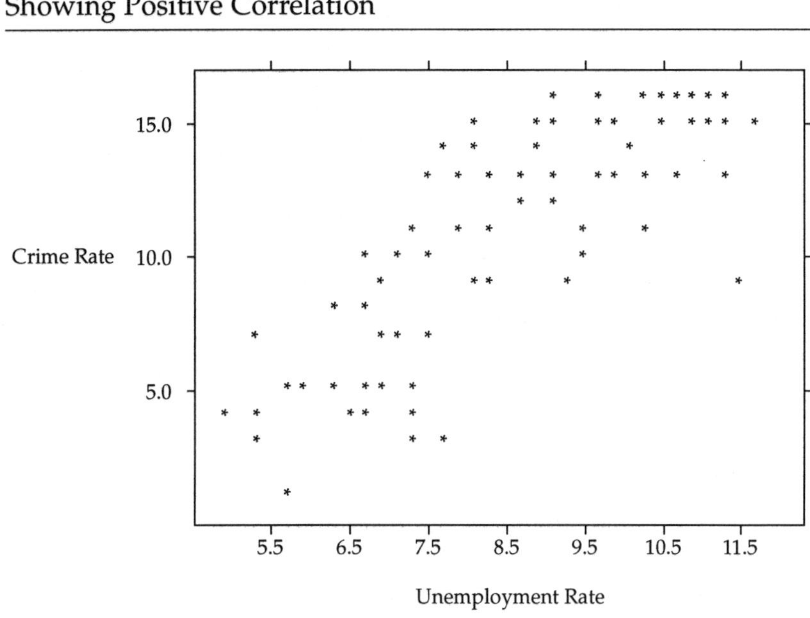

rates, as depicted with hypothetical data on cities in Figure 6.1. In this figure, the correlation is described as a **positive correlation** because high unemployment rates go with high crime rates and low unemployment rates go with low crime rates.

Some general patterns of covariation display **negative correlations**. If people who work in *less* bureaucratic settings display, on the average, *more* job satisfaction than people who work in more bureaucratic settings, then these two things, job satisfaction and degree of bureaucratization of work, are negatively correlated. This pattern can be depicted in a plot of employee data, as in Figure 6.2 which presents hypothetical evidence conforming to the stated pattern. According to the diagram, bureaucratization explains variation in job satisfaction because job satisfaction is high when people work in settings that are less bureaucratized, and vice versa.

In both examples, features of cases, called variables, are observed not in the context of individual cases, but *across* many cases. It is the pattern across many cases that defines the relation between the two features, not how the two features fit together or relate in individual cases. In the example of the positive correlation just described, it may be that one of the

FIGURE 6.2

Plot of Job Satisfaction and Bureaucratization of Work Showing Negative Correlation

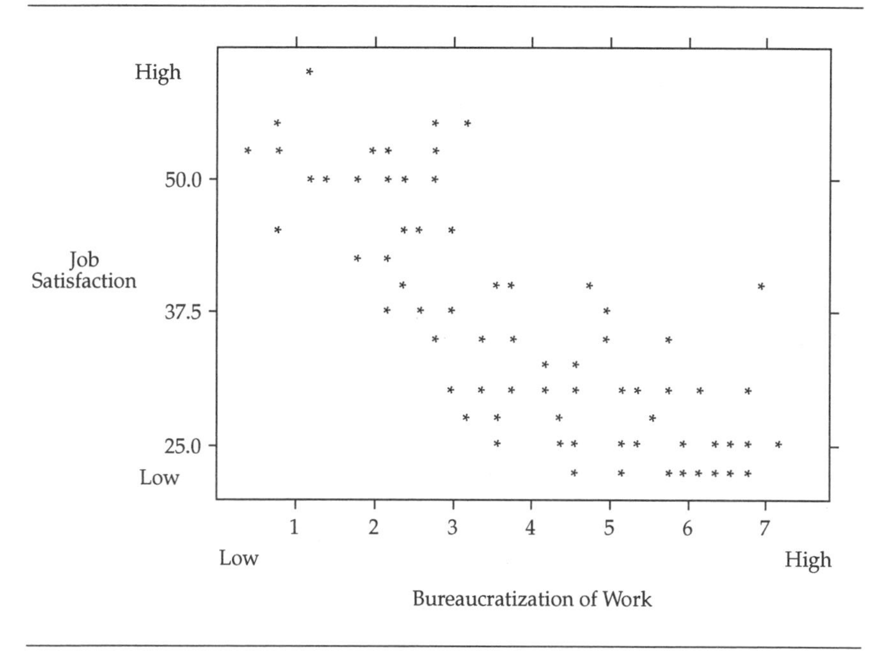

cities combining high unemployment and high crime rates had a recent, dramatic increase in unemployment coupled with a decrease in its crime rate—the opposite of the general pattern across cities. (If this city's crime declined from a very high level to a merely high level, it would still appear in the high unemployment–high crime rate portion of Figure 6.1.) What happened in one case over time cannot be addressed in the correlation across many cities at a single point in time. What matters is the general pattern: Do the cities with the highest unemployment rates also have the highest crime rates? In other words, the analysis of the relation between unemployment and crime in this example proceeds across cities, not within individual cities over time.

The correlation coefficient provides a way to make a direct, quantitative evaluation of the degree to which phenomena (for example, unemployment rates and crime rates) covary across cases (such as cities in the United States). The Pearson correlation coefficient itself varies between –1.00 and +1.00. A value of –1.00 indicates a perfect negative correlation; a value of +1.00 indicates a perfect positive correlation; and a value of 0 indicates no correlation. Sometimes a finding of no correlation is important because social researchers may have strong reasons to believe that a correlation should exist. The finding of no correlation may challenge widely accepted ideas.

It is sometimes difficult to specify what value constitutes a "strong" correlation. People tend to be relatively unpredictable. Thus, some researchers consider an individual-level correlation strong if it is greater than .3 (or more negative than –.3). For whole countries, by contrast, a correlation of .3 is considered weak because many features of countries tend to be highly correlated (for example, average wealth, life expectancy, literacy, level of industrialization, rate of car ownership, and so on). When assessing the strength of correlations, it is important to consider the nature of the data used in the computation.

Computing Correlation Coefficients

The hand calculation of a correlation coefficient is time consuming but straightforward. Usually, computers are used to compute correlation coefficients such as Pearson's *r*. The calculation of Pearson's *r* is illustrated in the appendix to this book in order to show the underlying logic of the coefficient.

Remember, the goal of the computation is to assess the degree to which the values (or scores) of two variables covary across many cases, in either a positive or a negative direction. In other words, do the cases with high values on the independent variable tend to have high values on the dependent variable? Do the cases with low values on the independent variable tend to have low values on the dependent variable? If so, then a strong positive correlation exists. If high values on the independent variable tend to be associated with low values on the dependent variable, and vice versa, then a strong negative correlation exists. If there is no pattern of covariation between two variables, then there is no correlation between them.

The key to calculating a correlation coefficient is to convert the scores on two variables to Z scores, as explained in the appendix. Z scores standardize variables so that they all have the same mean or average value (0) and the same degree of variation. Table 6.1 reports data on two variables for forty countries: the average number of calories consumed per person each day (the independent variable) and life expectancy (the dependent variable). These two variables can be used to test the simple idea that in countries where nutrition is better (as reflected in more calories consumed per person) people tend to live longer (as indicated in a

TABLE 6.1

Calculating the Correlation between Calorie Consumption and Life Expectancy

Country	*Life Expectancy*	*Calorie Consumption*	*Life Expectancy Z Scores*	*Calorie Consumption Z Scores*
Niger	45	2432	–2.04	–.70
Ethiopia	47	1749	–1.85	–1.92
Mali	47	2074	–1.85	–1.34
Uganda	48	2344	–1.75	–.86
Senegal	48	2350	–1.75	–.85
Sudan	50	2208	–1.55	–1.10
Ghana	54	1759	–1.17	–1.90
Kenya	58	2060	–.78	–1.37
Zimbabwe	58	2132	–.78	–1.24
Botswana	59	2201	–.68	–1.11
Indonesia	60	2579	–.58	–.44
Morocco	61	2915	–.49	.16
Peru	61	2246	–.49	–1.03
Philippines	63	2372	–.29	–.81
Thailand	64	2331	–.19	–.88
Turkey	64	3229	–.19	.72
Syria	65	3260	–.10	.77
Brazil	65	2656	–.10	–.30
Colombia	66	2543	.00	–.50
Paraguay	67	2853	.10	.05
Mexico	69	3132	.29	.55
S. Korea	69	2907	.29	.15
Malaysia	70	2730	.39	–.17
Hungary	70	3569	.39	1.33
Poland	71	3336	.49	.91
Chile	72	2579	.58	–.44
Jamaica	74	2590	.78	–.42
Ireland	74	3632	.78	1.44
United States	75	3645	.87	1.46
Greece	76	3688	.97	1.54
Australia	76	3326	.97	.89
Spain	77	3359	1.07	.95
Italy	77	3523	1.07	1.24
Netherlands	77	3326	1.07	.89
France	77	3336	1.07	.91
Canada	77	3462	1.07	1.14
Sweden	77	3064	1.07	.43
Norway	77	3223	1.07	.71
Switzerland	77	3437	1.07	1.09
Japan	78	2864	1.17	.07

longer life expectancy). Table 6.1 also reports the Z scores for these two variables for all forty cases.

Notice that countries with high scores on life expectancy have positive scores on life expectancy Z scores, and countries with low scores on life expectancy have negative scores on life expectancy Z scores. The same is true for calorie consumption. When the Z scores for two variables are multiplied, the products indicate a lot about the correlation. If high scores on one variable correspond to high scores on the other, and low scores on one correspond to low scores on the other, then the products of the Z scores will usually be positive, indicating a positive correlation. However, if low scores on one variable generally correspond to high scores on the other, and vice versa, then the products of the Z scores generally will be negative, indicating a negative correlation.

As the appendix illustrates, when the products of pairs of Z scores for two variables are averaged over all the cases, the number that results is Pearson's correlation coefficient, a number which varies between –1.00 (perfect negative correlation) and +1.00 (perfect positive correlation). The correlation between life expectancy and calorie consumption for the forty countries in Table 6.1 is .802, a strong positive correlation. The strong covariation between these two variables is clear from simply examining the table because the countries are sorted according to their values on life expectancy. The calculation of the correlation coefficient provides a direct, quantitative assessment of the degree to which the two measures covary.

Using Correlation Coefficients

The most basic use of correlation coefficients is to assess the strength of the relation between two variables. The correlation between calorie consumption and life expectancy is strong (r = .802), suggesting that an important key to longer life expectancy is nutrition. But there are many other uses of correlations. Most of these involve the comparison of competing causes, as indicated in the strength of correlations.

Consider the correlations reported in Table 6.2. The table shows all the correlations among four variables: three independent variables (calorie consumption, GNP per capita, and doctors per capita) and one dependent variable (life expectancy). (Notice that a variable correlates perfectly with itself, as shown by the values of 1.000 in Table 6.2.) GNP per capita is a rough measure of the wealth of a country. Doctors per capita is a rough measure of the availability of medical care.

TABLE 6.2

A Correlation Matrix with Three Independent Variables and a Dependent Variable

	Dependent Variable	*Independent Variables*		
	Life Expectancy	*Calorie Consumption*	*GNP per Capita (US$)*	*Doctors per Capita*
Life expectancy	1.000	.802	.651	.721
Calorie consumption	.802	1.000	.848	.321
GNP per capita (US$)	.651	.848	1.000	.671
Doctors per capita	.721	.321	.671	1.000

The first column shows the correlations of the three independent variables with the dependent variable. Calorie consumption is the most strongly correlated with life expectancy (r = .802), followed by doctors per capita (r = .721), followed by GNP per capita (r = .651). Is it possible to conclude from this evidence that all that really matters for life expectancy is calorie consumption? In other words, if the goal is to understand the variation in life expectancy across countries, is knowing nutrition levels enough? Is it reasonable to ignore the correlations with GNP per capita and doctors per capita?

In order to answer a question like this, it is not enough simply to identify the independent variable with the strongest correlation with the dependent variable. It is also necessary to examine the correlations among the independent variables. Consider first the correlation between calorie consumption and GNP per capita. It is strong (r = .848), suggesting that countries with the best nutrition are also the richest. Given that (1) these two independent variables are strongly correlated and (2) calorie consumption has a stronger correlation with life expectancy than does GNP per capita (r = .802 versus .651), it is reasonable to conclude that the link between calorie consumption and life expectancy is more fundamental than the link between GNP per capita and life expectancy. In short, richer countries have better nutrition, but it is good nutrition that causes greater life expectancy, not wealth per se.

What about doctors per capita? The correlation between doctors per capita and calorie consumption is positive, but not strong (r = .321). Thus, in some countries nutrition may not be good, but good health care is available, while in other countries, the opposite may be the case. In other words, doctors per capita and calorie consumption are not closely linked across countries in the same way that GNP per capita and calorie consumption are. Thus, the correlation between doctors per capita and life expectancy, the dependent variable, is relatively independent of and separate from the correlation between calorie consumption and life expectancy. Even though the correlation between doctors per capita and life expectancy (r = .721) is not as strong as the correlation between calorie consumption and life expectancy (r = .802), it is an important correlation. The pattern of correlations in Table 6.2 indicates that both doctors per capita and calorie consumption affect life expectancy.

A lot can be learned from looking at a correlation matrix like the one in Table 6.2. However, some quantitative studies examine many independent and dependent variables. Quantitative researchers use advanced statistical techniques such as multiple regression analysis to disentangle correlations among independent variables and assess their separate effects on dependent variables. They also use exploratory data analysis techniques ("EDA"; see Tukey 1977) to go beyond broad patterns of covariation to identify sets of cases that deviate from these broad patterns or to uncover very subtle patterns. Sometimes these techniques can be used to identify complex patterns of causation that are specific to subsets of cases included in a study (Leamer 1978). These advanced statistical techniques are very powerful data techniques and they further the primary goals of this approach: assessing general patterns (including their limits), making projections about the future, and evaluating broad theories.

Conclusion

Quantitative methods are best suited for addressing differences across a large number of cases. These methods focus especially on the covariation between attributes that vary by level, usually across many cases. If two features of cases vary together in a systematic way, they are said to correlate. Correlation is important because it may suggest that a causal or some other kind of important relation exists between the two features that are linked. Quantitative methods provide a direct way to implement a researcher's interest in general patterns, and quantitative researchers believe that these patterns of covariation provide important clues about social life.

In many ways, the quantitative approach appears to be the most scientific of the three approaches presented in this book. It favors generality and parsimony. It uses generic units such as individuals, families, states, cities, and countries. It can be used to assess broad relationships across countless cases. It condenses evidence to simple coefficients, using mathematical procedures. It can be used to test broad theoretical arguments and to make projections about the future. In short, it imitates many of the features and practices of hard sciences such as physics and chemistry.

While the quantitative approach does have many of the features of a hard science, it would be a mistake to portray this approach as something radically different from the other two strategies. All social research engages theoretical ideas and analytic frames, at least indirectly. All social research involves constructing images from evidence, usually lots of it. And all social researchers construct images by connecting social phenomena.

AFTERWORD

The Promise of Social Research

with Mary Driscoll

The Unity and Diversity of Method

Social scientists study and represent social life in many different ways. Sometimes they try to see social life through the eyes of the people they study. Sometimes they reconstruct significant historical events and pinpoint the relevance of these events to who we are. Sometimes they uncover broad patterns linking social phenomena across many cases and assess the implications of these patterns. And sometimes they try to map the diverse ways that social life is organized and practiced.

Although social scientific representations vary, they are especially well suited to the task of generating useful knowledge about social life because they (1) address phenomena that are socially significant, (2) link these phenomena to social theory and other thinking about social life, (3) incorporate large amounts of purposefully collected evidence, and (4) are constructed from systematic analyses of such evidence. In general, social researchers believe that this basic "formula" for constructing representations results in representations of social life that offer the greatest insight into its fundamental character. The promise of social research lies in the strength of this formula for constructing representations.

The three ways of linking ideas and evidence discussed in Part II of this book illustrate different ways of constructing representations following these general guidelines. Although these research strategies are only three among the many strategies that social researchers use, they illustrate the wide range of possible approaches. What these strategies share is a concern for grounding representations in evidence and ideas. They differ in how this grounding is achieved. In each of the following studies, ideas and evidence are linked in different ways. The studies illustrate how strategies can be matched to goals to produce empirically grounded, theoretically informed representations.

The quantitative study of covariation looks across cases and attempts to show how features of cases vary together (see Chapter 6). For example, in *Village Republics,* Robert Wade (1988) studied agricultural villages in southern India and showed a systematic correspondence between water supply (the villages used the same irrigation canals) and the degree to which each village developed institutions to cope with village-wide issues. The pattern he found was straightforward: the more serious the problem of water scarcity, the greater the development of village institutions of self-government. Villages at the head of the irrigation canal had ample water supply and developed few institutions. Villages at the end of the canal faced serious water shortages and developed elaborate institutions of self-government. These institutions were important because a whole village could suffer if any single farmer took too much irrigation water. The self-government of villages situated near the middle of the irrigation canal was less elaborate than that of the villages at the end of the canal, but more elaborate than that of the villages at the head of the canal.

Representations of social life that are based on patterns of covariation, such as Wade's, are grounded in evidence showing that there is a systematic relationship between variables. Two or more phenomena (such as water scarcity and institutions of self-government) parallel each other across many cases (villages). In Wade's study, the examination of covariation made it possible to show the ecological bases of village institutions.

Comparative studies of diversity, by contrast, pinpoint the differences among cases that separate them into distinct types (see Chapter 5). Generally, researchers who use this strategy attempt to identify *similarities* among the cases *within* a given type and *differences between* cases conforming to different types. Daniel Chirot (1994), for example, studied modern tyrants and showed that they fall into two major types—ideological and traditional. As it turns out, Chirot found that ideological tyrants (that is, those like Hitler and Stalin who tried to use their social and political ideas and aims to create a specific type of society) caused much more human suffering than did traditional tyrants. In fact, Chirot identified a number of factors, in addition to amount of suffering, that separate ideological and traditional tyrants.

Representations that are based on the use of contrasts to elaborate different types, such as Chirot's, are strongly grounded in evidence because the investigator identifies *clusters* of differences. By inspecting a range of different cases, investigators show that the differences separating cases into types are linked together empirically. For Chirot, this involved, among other things, showing the link between the content of the ideologies held by tyrants and the nature of the suffering they inflicted.

The qualitative study of commonalities reveals what the members of a group or category share (see Chapter 4). Mary Driscoll (1993), for example, studied women with food and body issues, including women who "shed" secondary sex characteristics by starving themselves and women who "take on size" by overeating. She also studied other groups, such as women who change their appearance through extreme covering with clothing and those who transform their apparent sex by dressing and passing as men. The commonality that unites these different groups is indirect—bodily resistance to gender prescriptions—detected by Driscoll through observation, in-depth interviews, and comparisons with studies of other groups.

Representations that are based on commonalities, such as Driscoll's representation of women with food and body issues, are strongly grounded in evidence because the investigator carefully pinpoints features that are consistent across cases. As explained in Chapter 4, this strategy achieves a solid grounding in evidence through in-depth study of many aspects of a relatively small number of cases. The search for commonalities continues until the researcher is confident that relevant commonalities have been identified and linked conceptually, and individual cases that deviate from the rest have been accounted for. In Driscoll's study, the search for commonalities made it possible for her to show that practices that appear to be very different on the surface (for example, starving, overeating, extreme cover, and cross-dressing) are all body transformations that are responses to gender limitations.

In a given investigation, most social researchers follow a single approach because the goals of the study usually dictate a particular strategy. For example, a researcher who wants to explain how crime rates vary across neighborhoods will use quantitative methods to examine the correlates of these rates. Some researchers, however, have goals that require the use of several analytic strategies in the same project. Arlie Hochschild, for example, used all three strategies just described in her study of working parents, *The Second Shift* (1989).

Hochschild based her book on interviews and observations of fifty working couples with children, their neighbors and friends, and their children's teachers, day-care workers, and babysitters. Altogether, Hochschild and her research associates interviewed 145 people. What typically happens to housework and child care when both parents work? Who takes responsibility for which tasks? What different kinds of arrangements do couples negotiate? How do couples differ? In answering these questions, Hochschild identified *commonalities* across the fifty working couples, explored their *diversity,* and revealed important patterns of *covariation.*

One fundamental *commonality* that Hochschild identified across the fifty families is captured in the title of the book—*The Second Shift.* In the fifty families she studied Hochschild found that almost universally the wives, not the husbands, worked a "second shift" of housework and child care after completing a full day of work in the paid labor force. She calculated that, on average, this second shift added an extra month of 24-hour workdays each year for the women. In the relatively few families where the work of the second shift was shared equally by the husbands, the wives still carried more of the emotional load and felt more responsible for the home front—worrying about the welfare of the children and the household and feeling torn between work responsibilities and the needs of family members.

Hochschild also explored the *diversity* of ways in which working couples cope with and adapt to the demands of two jobs, home, and family. Based on her examination of similarities and differences across the fifty families in her study, she specified a variety of common arrangements and accommodations. She used the orientations of the husbands and the wives (for example, traditional versus egalitarian) and their different ways of dividing labor in the home to sort the fifty families into ten different types. By specifying these different types and presenting representative cases, Hochschild was able to show the different strains and conflicts that accompany different adaptations to the pressures experienced by working parents.

Finally, Hochschild explored systematic patterns of *covariation* across the fifty families. In addition to documenting the inequality in the domestic workloads of wives and husbands, especially the average number of hours they devote to specific tasks, Hochschild also examined more subtle relationships. For example, she examined the degree to which the ideologies of the husbands regarding helping out at home (that is, their expressed attitudes about sharing this work) corresponded with their actual behaviors. She found that husbands with a more traditional orientation helped out less than did those with a more egalitarian orientation: 22 percent of husbands with a traditional orientation shared child care and housework equally with their wives, whereas 70 percent of husbands with an egalitarian orientation shared these tasks equally. By examining patterns of correspondence, Hochschild was able to address general questions about differences among the fifty families.

Hochschild used several analytic strategies because her study of working couples addressed many different aspects of the phenomenon.

By identifying commonalities, exploring diversity, and examining covariation, she constructed a rich portrait of working parents, the pressures they face, and their different ways of coping with these pressures.

Diverse ways of studying and representing social life exist because social researchers have many different goals. These goals range from broad objectives that are common to many types of scientific research, such as identifying general patterns and testing theories, to goals that are more specific to *social* research, such as interpreting historical events and giving voice to specific groups (see Chapter 2). While the goals of social research are diverse, there is unity in this diversity. All good social research contributes to knowledge of social life, which in turn provides important keys to understanding who we are, to comprehending social diversity, to addressing the root causes of social phenomena of general or public concern, and to anticipating future patterns and trends.

Social Research and Its Critics

While social scientific representations of social life derive strength from their grounding in strategically chosen ideas and evidence, they are necessarily partial and imperfect. Every representation is open to criticism, and it is important to address some common criticisms of social research.

Uniqueness

One common complaint about social research is based on the humanistic idea that every person and situation is unique and should be understood in its uniqueness. This thinking sees social scientific categories and variables as crude and cumbersome tools that can do more harm than good by lumping people together, generalizing about them, and ignoring their individuality.

It is true that social research focuses on broad rather than particular understandings. Social researchers do not reject the idea that everything should be appreciated in its uniqueness. But the appreciation of uniqueness is simply not the task of social science. Social research is driven by a focus on understandings that extend beyond the singular, beyond our own selves and intimate others, and to the many—to groups, peoples, and societies. The individual case is woven into larger social patterns so that more general knowledge can be gained. Even studies of a single individual (for example, Harper 1987) are driven by an interest in what one case can teach us about ourselves and others—something larger than that one person.

We have much to gain from connecting the particular (data collected about a particular person, place, or event) with other accumulated data and with the general thinking and research on a topic. By interpreting the individual case as an elaboration of something more universal, social scientists can draw on a large body of ideas and previous research. This knowledge offers the discriminating researcher a valuable resource of accumulated wisdom as well as points of reference for comparison and analysis. The connection of individual cases to data sets and of data sets to abstract, long-standing debates about social life (for example, about the causes and consequences of social inequality) anchors representations, tying them not only to the community of researchers, but also to the concerns of the informed public.

What social research loses in not representing the unique person, place, or event, it strives to gain in representing social affairs in general. One of the main contributions of social research is to identify the social threads that run through individual problems, showing when they are in fact instances of massive trends. Consider a few examples.

Absenteeism, accidents on the job, and job turnover may increase among factory workers, even when factory jobs are scarce. Social researchers suspect there are important patterned responses in what may appear to be the unrelated, individual acts of workers. They may find that the increase in employee problems is in fact a reaction to work speedups and other changes caused by industry efforts to accelerate productivity in the face of international competition.

Another example: Many adults assume that differences in intelligence are the primary reason one child thrives in school while another does poorly. Various studies addressing variation in academic performance pinpoint key social factors that influence concentration, learning, and test performance, far more than do IQ differences. This knowledge can be used to help children learn.

Consider this example: Parents may be puzzled to find that their children seem especially prone to violence. Research on violence and the media helps them make informed decisions about the possible link between the upsurge in violent media images and their children's violent behavior.

Still another example: Parents and doctors struggle with a young girl who refuses to eat and who is literally starving to death. They may think that her problem lies solely within her individual psyche. It may benefit all concerned to learn that eating disorders, which have reached startling levels among young girls and women in the 1980s and 1990s, mimic gendered illnesses from other historical periods (such as hysteria in the middle to late 1800s and agoraphobia in the 1950s and 1960s). Scholars interpret epidemic levels of illness among females as constituting, in part, an unconscious protest against limiting gender roles. Knowledge of these trends may enlighten both caretakers and those who are stuck in self-defeating behaviors.

As these examples show, social scientists link the individual with society, and in doing so, extend the personal circumstances of everyday life into the realm of social issues. C. Wright Mills (1959) explained in his book *The Sociological Imagination* that it is the task of social scientists to take the "private troubles" that people experience, often in isolation from one another, and show their connection to "public issues." Researchers study and represent both the individual and the larger milieu in concert. The particular and the general inform each other, and good social scientific representations increase our understanding of both. In short, good social research helps us understand what is going on in the world and in ourselves.

Thus, as we consider what is lost when social and human characteristics are categorized, conceptualized, or correlated as data, we must also consider what is gained. The individual life is illuminated with general knowledge of social life. Social research lifts the individual into historical and cultural perspective.

The focus on the larger social order, versus the individual, has additional gains. In order to make predictions about the future, social scientists must rely on the broad, accumulated knowledge that links their cases with larger patterns. Knowledge of general patterns makes it possible to extrapolate future trends and possibilities. For example, what we know about crime, lifestyles and consumption, family patterns, job opportunities, and ethnic and racial conflict allows us to make useful predictions about them.

Another gain comes from the study of societies and settings that differ from our own. Research of this type may challenge conventional thinking because it often shows that what is normative and accepted as the "ways things are" is relative to time and place. Research on social milieus that differ from our own opens up new avenues of thinking and being. Alternative ways of addressing common social issues become possible. Differences between people are understood and accepted, and mutuality is increased. Thus, through an examination of similarities and differences, social research connects the individual to society and societies to each other. Without connecting the particular and the general, the "social" would be missing from social research.

Multiplicity

The tension between representing social life in particular and representing it in general is inherent in the practice of social research. This tension within social research is often the source of additional criticisms of social scientific representations.

Consider the problem of representing the differences that exist within any social category. The lives of homeless people, for example, are complex and diverse. The world of a homeless, African-American woman who lives with her children in the inner city is vastly different from that of a homeless, single, white man who lives alone in a rural area. A researcher might try to study homelessness in an encompassing way, but it would be difficult to offer an understanding that embraces the worlds of these two people. Even the researcher who attempts to take race, age, gender, and other important defining factors into account will find innumerable other factors that differentiate the lives of homeless people. Thus, social scientific representations are sometimes criticized for failing to address important differences in an encompassing way.

On top of this, social scientific representations of a *single* situation, place, or event can diverge sharply. As a research subject, the same homeless person may be presented as a victim by one researcher and as a schemer by another. Another study might represent him as a mentally ill person. The fact that it is possible to represent the same subject of research in various ways casts doubts on the claim that social scientific representations derive strength from their grounding in evidence.

Yet, a single, fixed picture of homelessness, or of any other topic that touches social life, would be a false one. A science that demands a clear, final portrait of phenomena could not be a *social* science. Human variation and the fluidity and open-endedness of social life require a science that is not static or fixed. These features of social life demand a science that can capture, with clarity, what is going on in the world and with people, while at the same time representing a good deal of its diversity—a variety that is endless and ever changing. Of necessity, the social sciences must make room for imprecision and incompleteness in the study of human affairs.

Thus, the practice of social research tests researchers in many ways. How does a social scientist capture a meaningful picture of human ingenuity under oppression, the role of computers in the globalization of cultures, the collapse of Soviet-style communism in Eastern Europe, or the changing meaning of aging and dying across several generations in a small town? It is not enough for social scientists to bring their intellects to bear on the questions before them. They must bring their imaginations as well.

Many researchers talk about how they work back and forth between a rational research plan, on the one hand, and hunches, intuitions, vague notions, gut reactions, brainstorming sessions, and other ways of stoking the imagination, on the other. The toil of doing rigorous, grounded research is often aided by leaps of the imagination, which lead to breakthroughs and new discoveries. Some social researchers like to think of themselves not simply as scientists, but also as artists or craftspersons. Their final products, the representations they construct, may embody both grounded scientific effort and creative flight.

Also, depending on the nature of the investigator's questions and strategies, often a researcher's heart is deeply engaged in his or her research. In fact, a compassionate heart guides some researchers through their research projects as much as an engaged mind does. The idea that a researcher's deep emotional and political investments may offer guidance contradicts the hard science notion of doing objective, unbiased research. The study of human affairs is, however, by necessity an artful science.

Social Research: A Collaborative Journey

The indefinite nature of studying human affairs and the impossibility of getting a "real take" on it lead some to conclude that social life cannot be represented in a valid way. In this view, all things, including life's problems, are mere "social constructions." It is true that our understandings of social life are based on somewhat subjective and variable interpretations of evidence that is laden with many meanings. However, these problems do not pose insurmountable challenges to social research.

First of all, the variability of social life does not change the fact that groups of people share large pools of knowledge about the world. We have common, agreed-upon meanings for many things in life, from how to cross a busy street safely to how to interpret subtle cues in intimate relationships. Some of these understandings are informal and unconscious (for example, many aspects of gender socialization), and others are explicitly learned (how to drive a car or how to groom oneself for job advancement). In order to get along in life, and even more, to prosper, we rely on myriad, shared understandings of how the world works. Social research adds to our common stock of knowledge about life. It helps us understand the meaning of our world and larger historical processes and events.

The representations that convey these meanings are not perfectly precise or complete, but they embody the negotiated understandings of a community of scholars who review each other's work and who collectively, through time, determine which representations have the best basis in ideas and evidence. In essence, the multiplicity of the social world—the fact that the same evidence can be interpreted in many ways—is made manageable by the social scientific community, which decides which views of social life make the most sense. Just as the individual case becomes one within a chorus of cases, most social research represents a chorus of researchers. Social research is strengthened and clarified by scholars who contribute to one another's projects before and after they are published. The validity of social research lies in its contributions to common knowledge about the world, knowledge that is continuously revised and updated by the social scientific community.

It is reassuring to know that the representations that social scientists construct are accountable to a community of scholars, that they are grounded in carefully chosen ideas and evidence, and that they are informed by rigorous and well-tested theories and methods. Still, it should be clear that no study on a topic can be considered definitive. Different studies illuminate different aspects of a single topic. For example, images of the homeless person as victim, schemer, or mentally ill may be based on valid interpretations of different data collected on one person combined with different ideas about those data. Each representation may be valid social science, and each may add to our understanding of homelessness as a social phenomenon. We might think of each study of homelessness, or of any other social phenomenon, as a piece in a puzzle or as a patch in a quilt. Researchers who study homelessness each contribute a piece toward a fuller understanding of the problem, and each piece may contribute as well to solutions. Social research is a collaborative journey that brings together many people with different combinations of ideas and evidence and thus many different representations. These efforts, taken together as a whole, can far exceed their sum.

To conclude that we cannot understand social life because it is complex, shifting, and variously interpretable is tantamount to giving up on the idea that we can increase our understanding of human affairs and use this knowledge to make things better. This thinking may lead some to stop trying to understand others or feel concern for the common good. The fact that homelessness can be represented in a variety of ways, each imperfect and biased, does not mean homelessness is not a real problem, nor does it absolve our collective responsibility for it.

We live in an age of social scientific reasoning, in which the results of social research are filtered to the general public through bookstores, textbooks and school systems, governments and social agencies, and the media. The increasing importance of social science for public audiences can be seen clearly in widespread public views and media treatment of social issues. Thirty years ago most Americans would flatly reject the idea that many poor people are poor because there are not enough good jobs to go around. Instead, they would claim that poor people are poor because they are lazy. Today, most Americans would agree with the results of decades of social research showing that poverty has important social causes, not just individual ones.

In an age when information is quickly and easily disseminated, the representations of social life that social researchers construct take on a new significance. These representations often reach many people quickly and contribute to their views on many topics. They affect social policy and may have a powerful impact on popular discussion of social issues, even when this impact is indirect. Thus, the results of social research are more accessible and perhaps more relevant than ever to people grappling with a complex, rapidly changing world.

This afterword has addressed several criticisms of social science. The problem of representing the diversity within social life and the problem of representing it adequately when there are so many ways to approach the same phenomenon are both real problems. These and other issues that challenge our capacity to conduct good social research today are at the core of its current task. They also reflect the challenges that people face in today's world. The inherent tension between understanding the particular and the general in social life is as unresolved in ourselves as a people as it is in the social sciences. Consider how we as a collectivity are confronted by questions about human uniqueness and how to respect essential rights. On the national level alone, we are immersed in issues of class, gender, race, and ethnicity; issues of democratic power and how much government we should have; conflicts over how to use versus how to protect the environment; and ethical issues centered on the body—abortion, euthanasia, reproduction technologies, cosmetic surgery, and the allocation of health care, to name only a few. These and other issues of freedom and responsibility abound.

How do we understand the great diversity within human affairs? How do people take into account a fast-changing world that connects societies and cultures more closely than ever before—through faster forms of travel and expanded migration, the media, new forms of technology and trade, international relations, and the globalization of markets?

Social research reaches the end of this century with more than a hundred years of published research. Yet, it is not enough to face the twenty-first century with this stock of knowledge. Social scientists must address the specific concerns and questions of every age. It will take the collaborative powers of many to meet the questions of the twenty-first century and to construct useful representations of social life for all who are curious about people, concerned about social issues, and committed to the quality of human life.

Computing Correlation Coefficients

The first step on the road to computing the correlation coefficient is the computation of the mean (or average) level of the independent and dependent variables. In this example, refer to Table A.1, which reports data on two variables for forty countries: the average number of calories consumed per person each day (the independent variable) and life expectancy (the dependent variable). These two variables can be used to test the simple idea that in countries where nutrition is better (as reflected in more calories consumed per person) people tend to live longer (as indicated in a longer life expectancy).

It is first necessary to compute the mean level of calorie consumption per capita and life expectancy across the forty countries so that it is possible to determine which values are high and which are low. Values that are well above the mean are considered high; values well below the mean value are considered low.

The computation of the mean of a variable is straightforward. Simply sum the values for all the cases, and then divide by the number of cases. The formula for the mean is

$$\overline{X} = \frac{\sum x_i}{N}$$

where $\overline{X}$ is the symbol for the mean of the variable; $\sum$ indicates that the values are summed; x_i indicates the actual values of the variable to be summed (in our example the forty country scores), and N is the number of cases (forty). The mean life expectancy for the forty countries in Table A.1 is 66 (2640/40; see column 1); the mean calorie consumption is approximately 2825.52 (113,021/40; see column 3). (The tables in this appendix contain a good deal of rounding error because of limitations on the number of decimal places it is reasonable to report.)

The next step is to assess the degree to which cases are above or below the mean on the two variables. To do this researchers use **deviation scores**. To compute a deviation score, subtract the mean value of a

TABLE A.1

Calculating the Covariance of Calorie Consumption and Life Expectancy

	1	2	3	4	5	6
Country	*Life Expectancy*	*Life Expectancy Deviations*	*Calorie Consumption*	*Calorie Consumption Deviations*	*Column 2 × Column 4*	*Column 2 Squared*
Niger	45	–21	2432	–393.52	8263.92	441.00
Ethiopia	47	–19	1749	–1076.52	20453.88	361.00
Mali	47	–19	2074	–751.52	14278.88	361.00
Uganda	48	–18	2344	–481.52	8667.36	324.00
Senegal	48	–18	2350	–475.52	8559.36	324.00
Sudan	50	–16	2208	–617.52	9880.32	256.00
Ghana	54	–12	1759	–1066.52	12798.24	144.00
Kenya	58	–8	2060	–765.52	6124.16	64.00
Zimbabwe	58	–8	2132	–693.52	5548.16	64.00
Botswana	59	–7	2201	–624.52	4371.64	49.00
Indonesia	60	–6	2579	–246.52	1479.12	36.00
Morocco	61	–5	2915	89.48	–447.40	25.00
Peru	61	–5	2246	–579.52	2897.60	25.00
Philippines	63	–3	2372	–453.52	1360.56	9.00
Thailand	64	–2	2331	–494.52	989.04	4.00
Turkey	64	–2	3229	403.48	–806.96	4.00
Syria	65	–1	3260	434.48	–434.48	1.00
Brazil	65	–1	2656	–169.52	169.52	1.00
Colombia	66	0	2543	–282.52	.00	.00
Paraguay	67	1	2853	27.48	27.48	1.00
Mexico	69	3	3132	306.48	919.44	9.00
S. Korea	69	3	2907	81.48	244.44	9.00

TABLE A.1 (CONTINUED)

Calculating the Covariance of Calorie Consumption and Life Expectancy

	1	2	3	4	5	6
Country	*Life Expectancy*	*Life Expectancy Deviations*	*Calorie Consumption*	*Calorie Consumption Deviations*	*Column 2 × Column 4*	*Column 2 Squared*
Malaysia	70	4	2730	–95.52	–382.08	16.00
Hungary	70	4	3569	743.48	2973.92	16.00
Poland	71	5	3336	510.48	2552.40	25.00
Chile	72	6	2579	–246.52	–1479.12	36.00
Jamaica	74	8	2590	–235.52	–1884.16	64.00
Ireland	74	8	3632	806.48	6451.84	64.00
United States	75	9	3645	819.48	7375.32	81.00
Greece	76	10	3688	862.48	8624.80	100.00
Australia	76	10	3326	500.48	5004.80	100.00
Spain	77	11	3359	533.48	5868.28	121.00
Italy	77	11	3523	697.48	7672.28	121.00
Netherlands	77	11	3326	500.48	5505.28	121.00
France	77	11	3336	510.48	5615.28	121.00
Canada	77	11	3462	636.48	7001.28	121.00
Sweden	77	11	3064	238.48	2623.28	121.00
Norway	77	11	3223	397.48	4372.28	121.00
Switzerland	77	11	3437	611.48	6726.28	121.00
Japan	78	12	2864	38.48	461.76	144.00
SUM	2,640	0.00	113,021.00	0.00	180,428.00	4,126.00
SUM/N	66	0.00	2,825.52	0.00	4,510.70	103.15

variable (computed according to the simple formula just described) from each case's score on that variable:

$$x_i \text{ deviation score} = x_i - \overline{X}$$

A large positive result indicates that the case's score on the variable is well above the mean (a high value); a large negative result indicates that the case's score is well below the mean (a low value). Columns 2 and 4 of Table A.1 show the computation of deviation scores for all forty cases on both variables. For example, Niger, the first country in the table, has a deviation score on life expectancy of –21 (a life expectancy score of 45 minus the mean life expectancy of 66).

Notice that the countries in Table A.1 are listed according to their scores on the dependent variable, life expectancy (column 1 of the table). The countries with the lowest scores on life expectancy are at the top of the table; the countries with the highest scores are at the bottom. Thus, the deviation scores on life expectancy (column 2) range from large negative numbers at the top to large positive numbers at the bottom. Because life expectancy and calorie consumption scores covary, the deviation scores for calorie consumption (column 4) also tend to range from negative at the top of the table to positive at the bottom of the table. Of course, the computation of the correlation coefficient will provide an exact quantitative assessment of how closely these two variables covary.

One simple way to see if the scores of the two variables parallel each other in a positive or negative direction is to use their deviation scores (columns 2 and 4) to compute the **covariance** between the independent and the dependent variable. Covariance is the average of the products of deviation scores and is computed as follows:

$$\frac{\sum (x_i - \overline{X})(y_i - \overline{Y})}{N}$$

As the formula shows, for each case, the deviation score of the independent variable is multiplied by the deviation score of the dependent variable. After these products are calculated for each case, they are added together, and then this sum is divided by the number of cases. The result is the average product of the deviation scores (or covariance). The computation of the covariance between calorie consumption and life expectancy is shown in column 5 of Table A.1.

The computation of the covariance is very similar to the computation of the correlation coefficient, and it is useful to understand the strengths and weaknesses of the covariance before moving on to correlation. First, it should be noted that the sign of the covariance (positive or negative) is also the sign of the correlation. Notice that when the value of the independent variable is low, the sign of its deviation score is negative. The same is true for the dependent variable. When these low (that is, negative) deviation scores are multiplied, they result in a positive product. If low scores on one variable are generally matched with low scores on the other, and high scores are generally matched with high scores, then the sum of their products (as specified in the formula for covariance) will be a large positive number. A positive covariance indicates that the correlation between the two variables is also positive.

By contrast, if high scores on one variable are generally paired with low scores on the other, and vice versa, then their products will generally be negative. Adding these negative products together will result in a large negative number, indicating a negative correlation. Finally, if negative products and positive products balance each other out, then their sum (that is, their covariance) will be zero or close to zero, indicating that the two variables are not correlated.

Covariances tell us a lot, but they are awkward. In Table A.1, the covariance between calorie consumption and life expectancy is a very large positive number, indicating these two variables are positively correlated (that is, in countries where there is better nutrition, people tend to live longer). But it is difficult to determine the *strength* of the correlation from this large positive number. If calorie consumption, for example, had been measured in calories consumed per person per year instead of per day, the covariance would be a much larger number (by a factor of 365, the number of days in a year), but the degree of actual correspondence between calorie consumption and life expectancy would be unchanged.

The problem is to find a way to **standardize** variables so that the multiplication of their deviation scores is not affected by the size of their units (for example, calories per year versus calories per day). Ideally, this standardization should also produce a covariance that varies between –1.00 (indicating perfect negative correlation) and +1.00 (indicating perfect positive correlation). Fortunately, there is a way to standardize deviation scores so that the computation of the covariance results in just such a coefficient. Once variables are standardized, the computation of their covariance results in the Pearson correlation coefficient.

The best way to standardize a deviation score is to assess whether it is large or small relative to the size of the other deviation scores for a variable. Is a deviation score of 200 calories a large or a small positive deviation? This value must be compared to the typical deviation in order to make this assessment. There are several ways to compute the typical deviation. The most useful computation of the typical deviation is called the **standard deviation**, which is also indicated by the symbol σ. The standard deviation is computed as follows:

$$x \text{ standard deviation (or } \sigma_x) = \sqrt{\frac{\sum (x_i - \overline{X})^2}{N}}$$

Deviations from the mean are first squared (making them all positive values); then they are summed and divided by *N* (producing the average squared deviation); and then the square root of the average squared deviation is computed. Because the deviations are squared and then averaged, and then the square root of this average is computed, the result is consistent with the original units of the variable, not the squared units (for example, number of calories, not number of calories squared).

The computation of the standard deviation of life expectancy is shown in column 6 of Table A.1. This column shows what happens when life expectancy deviation scores (from column 2 of Table A.1) are squared. Notice that they are all positive values. These squared values are summed (shown at the bottom of column 6) and then divided by *N* (the number of cases, 40) to produce the average squared deviation, which for life expectancy is 103.15. Taking the square root of this number results in the standard deviation of life expectancy, 10.16. Parallel calculations on the deviation scores for calorie consumption result in a standard deviation for calorie consumption of 560.7 (again, these calculations reflect rounding error; computers give more exact figures).

Once the standard deviation is computed, it is possible to correct deviation scores in a way that makes them uniform in their units. This is a very important step in the calculation of the correlation coefficient. Examine Table A.2. Columns 1 and 3 show the deviation scores for life expectancy and calorie consumption taken from columns 2 and 4 of Table A.1. To produce standardized deviation scores (also known as "standard scores" and as *Z* scores), it is necessary simply to divide the deviation scores by the appropriate standard deviation. The relevant formula is

$$x_i \text{ standardized score (or } Z \text{ score)} = \frac{x_i - \overline{X}}{\sigma_x}$$

Column 2 shows standardized scores for life expectancy (its deviation scores have been divided by 10.16); column 4 shows standardized scores for calorie consumption (its deviation scores have been divided by 560.7). Notice that both sets of scores now range more or less between the same values. The highest standardized score for life expectancy is 1.17; the lowest is –2.04. The highest standardized score for calorie consumption is 1.54; the lowest is –1.92.

The next step in the computation of the correlation coefficient is to compute the covariance of the standardized scores. As it turns out, the covariance of standardized scores is the Pearson correlation coefficient. The formula is

$$\text{correlation coefficient (or } r) = \frac{\sum\left[\left(\frac{x_i - \overline{X}}{\sigma_x}\right)\cdot\left(\frac{y_i - \overline{Y}}{\sigma_y}\right)\right]}{N}$$

The Pearson correlation coefficient is a covariance that ranges from –1.00 to +1.00, and it equals 0 when there is no simple pattern of correspondence between two variables.

The computation of the correlation between life expectancy and calorie consumption is shown in column 5 of Table A.2. The values in column 2 are multiplied by the values in column 4 to produce the values in column 5. Notice that most of the products in column 5 of Table A.2 are positive. Thus, when this column is summed and the sum is divided by N, the result is a positive number. The last figure in column 5 is the correlation (r = .80), showing that there is a strong, positive relationship between calorie consumption and life expectancy. This finding indicates that in countries where nutrition is better, people live longer.

TABLE A.2

Calculating the Correlation between Life Expectancy and Calorie Consumption

	1	*2*	*3*	*4*	*5*
Country	*Life Expectancy Deviations*	*Life Expectancy Z Scores*	*Calorie Consumption Deviations*	*Calorie Consumption Z Scores*	*Column 2 × Column 4*
Niger	–21	–2.04	–393.52	–.70	1.43
Ethiopia	–19	–1.85	–1076.52	–1.92	3.55
Mali	–19	–1.85	–751.52	–1.34	2.47
Uganda	–18	–1.75	–481.52	–.86	1.50
Senegal	–18	–1.75	–475.52	–.85	1.48
Sudan	–16	–1.55	–617.52	–1.10	1.71
Ghana	–12	–1.17	–1066.52	–1.90	2.22
Kenya	–8	–.78	–765.52	–1.37	1.06
Zimbabwe	–8	–.78	–693.52	–1.24	.96
Botswana	–7	–.68	–624.52	–1.11	.76
Indonesia	–6	–.58	–246.52	–.44	.26
Morocco	–5	–.49	89.48	.16	–.08
Peru	–5	–.49	–579.52	–1.03	.50
Philippines	–3	–.29	–453.52	–.81	.24
Thailand	–2	–.19	–494.52	–.88	.17
Turkey	–2	–.19	403.48	.72	–.14
Syria	–1	–.10	434.48	.77	–.08
Brazil	–1	–.10	–169.52	–.30	.03
Colombia	0	.00	–282.52	–.50	.00
Paraguay	1	.10	27.48	.05	.00
Mexico	3	.29	306.48	.55	.16

TABLE A.2 (CONTINUED)

Calculating the Correlation between Life Expectancy and Calorie Consumption

	1	2	3	4	5
Country	*Life Expectancy Deviations*	*Life Expectancy Z Scores*	*Calorie Consumption Deviations*	*Calorie Consumption Z Scores*	*Column 2 × Column 4*
S. Korea	3	.29	81.48	.15	.04
Malaysia	4	.39	–95.52	–.17	–.07
Hungary	4	.39	743.48	1.33	.52
Poland	5	.49	510.48	.91	.44
Chile	6	.58	–246.52	–.44	–.26
Jamaica	8	.78	–235.52	–.42	–.33
Ireland	8	.78	806.48	1.44	1.12
United States	9	.87	819.48	1.46	1.28
Greece	10	.97	862.48	1.54	1.49
Australia	10	.97	500.48	.89	.87
Spain	11	1.07	533.48	.95	1.02
Italy	11	1.07	697.48	1.24	1.33
Netherlands	11	1.07	500.48	.89	.95
France	11	1.07	510.48	.91	.97
Canada	11	1.07	636.48	1.14	1.21
Sweden	11	1.07	238.48	.43	.45
Norway	11	1.07	397.48	.71	.76
Switzerland	11	1.07	611.48	1.09	1.17
Japan	12	1.17	38.48	.07	.08
SUM					32.09
SUM/*N*					.80

References

Baran, Paul. 1957. *The Political Economy of Growth*. New York: Monthly Review Press.

Barash, Meyer, and Alice Scourby. 1970. *Marriage and the Family*. New York: Random House.

Becker, Howard S. 1953. "Becoming a Marijuana User." *American Journal of Sociology* 59:235–42.

———. 1963. *Outsiders: Studies in the Sociology of Deviance*. New York: Free Press.

———. 1967. "Whose Side Are We On?" *Social Problems* 14:239–47.

———. 1986. "Telling about Society." In *Doing Things Together,* 121–36. Evanston: Northwestern University Press.

Becker, Howard S., Blanche Geer, Everett C. Hughes, and Anselm L. Strauss. 1961. *Boys in White: Student Culture in Medical School*. Chicago: University of Chicago Press.

Burawoy, Michael. 1979. *Manufacturing Consent: Changes in the Labor Process under Monopoly Capitalism*. Chicago: University of Chicago Press.

Chirot, Daniel. 1994. *Modern Tyrants: The Power and Prevalence of Evil in Our Age*. New York: Free Press.

Cressey, Donald R. 1953. *Other People's Money*. New York: Free Press.

Daniels, Arlene Kaplan. 1988. *Invisible Careers: Women Civic Leaders from the Volunteer World*. Chicago: University of Chicago Press.

Davis, James A., and Tom W. Smith. 1988. *General Social Surveys, 1972–1988: Cumulative Codebook*. Chicago: National Opinion Research Center.

Denzin, Norman. 1970. *The Research Act: A Theoretical Introduction to Sociological Methods*. Chicago: Aldine.

———. 1978. *Sociological Methods: A Sourcebook*. New York: McGraw Hill.

Diesing, Paul. 1971. *Patterns of Discovery in the Social Sciences*. Chicago: Aldine.

Drass, Kriss, and Charles C. Ragin. 1989. *QCA: Qualitative Comparative Analysis*. Evanston, Ill.: Center for Urban Affairs and Policy Research, Northwestern University.

Driscoll, Mary. 1993. "Margin Work: Women and Nonconformity in the Gender Margins." Unpublished manuscript. Department of Sociology, Northwestern University.

Dumont, Louis. 1970. *Homo Hierarchicus: The Cast System and Its Implications*. Chicago: University of Chicago Press.

Duneier, Mitchell. 1992. *Slim's Table: Race, Respectability, and Masculinity*. Chicago: University of Chicago.

Durkheim, Emile. 1951. *Suicide: A Study in Sociology*. New York: Free Press.

Ebaugh, Helen Rose Fuchs. 1977. *Out of the Cloister: A Study of Organizational Dilemmas*. Austin: University of Texas Press.

———. 1988. *Becoming an Ex: The Process of Role Exit*. Chicago: University of Chicago Press.

Eckstein, Harry. 1975. "Case Study and Theory in Political Science." Chapter 3 in *Handbook of Political Science*, edited by Fred I. Greenstein and Nelson W. Polsby. Reading Mass.: Addison-Wesley.

Esping-Andersen, Gosta. 1990. *The Three Worlds of Welfare Capitalism*. Princeton: Princeton University Press.

Evans, Peter. 1979. *Dependent Development: The Alliance of Multinational, State, and Local Capital in Brazil*. Princeton: Princeton University Press.

Feagin, Joe R., Anthony M. Orum, and Gideon Sjoberg. 1991. *A Case for the Case Study*. Chapel Hill: University of North Carolina Press.

Frank, Andre Gunder. 1967. *Capitalism and Underdevelopment in Latin America*. New York: Monthly Review Press.

———. 1969. *Latin America: Underdevelopment or Revolution*. New York: Monthly Review Press.

Glaser, Barney G., and Anselm L. Strauss. 1967. *The Discovery of Grounded Theory: Strategies for Qualitative Research*. London: Weidenfeld and Nicholson.

Goffman, Erving. 1961. *Asylums: Essays on the Social Situation of Mental Patients and Other Inmates*. Garden City, N.Y.: Anchor Books.

———. 1963. *Stigma: Notes on the Management of Spoiled Identity*. Englewood Cliffs, N.J.: Prentice-Hall.

Hanson, Norwood Russell. 1958. *Patterns of Discovery: An Inquiry into the Conceptual Foundations of Science*. Cambridge: Cambridge University Press.

Harper, Douglas. 1982. *Good Company*. Chicago: University of Chicago Press.

———. 1987. *Working Knowledge: Skill and Community in a Small Shop*. Chicago: University of Chicago Press.

Hochschild, Arlie. 1983. *The Managed Heart*. Berkeley: University of California Press.

Hochschild, Arlie (with Anne Machung). 1989. *The Second Shift: Working Parents and the Revolution at Home*. New York: Viking.

Hoover, Kenneth R. 1976. *The Elements of Social Scientific Thinking*. New York: St. Martin's Press.

Jenkins, J. Craig. 1983. "Resource Mobilization Theory and the Study of Social Movements." *Annual Review of Sociology* 9:527–553.

Katz, Jack. 1982. *Poor People's Lawyers in Transition*. New Brunswick, N.J.: Rutgers University Press.

Kuhn, Thomas. 1962. *The Structure of Scientific Revolutions*. Chicago: University of Chicago Press.

Lazarsfeld, Paul F., and Morris Rosenberg. 1955. *The Language of Social Research*. Glencoe, Ill.: Free Press.

Leamer, Edward. 1978. *Specification Searches: Ad Hoc Inference with Non-experimental Data*. New York: Wiley.

Lenin, Vladmir Ilyich. 1975. *Imperialism: The Highest Stage of Capitalism*. Moscow: Progress Publishers.

Lieberson, Stanley. 1985. *Making It Count: The Improvement of Social Research and Theory*. Berkeley: University of California Press.

Lijphart, Arend. 1971. "Comparative Politics and the Comparative Method." *American Political Science Review* 65: 682–93.

———. 1984. *Democracies: Patterns of Majoritarian and Consensus Government in Twenty-One Countries*. New Haven, Conn.: Yale University Press.

Lindesmith, Alfred R. 1947. *Opiate Addiction*. Bloomington: Principia Press.

Lipset, Seymour Martin. 1982. *Political Man: The Social Bases of Politics*. Baltimore: Johns Hopkins University Press.

Lyng, Stephen. 1990. "Edgework: A Social Psychological Analysis of Voluntary Risk Taking." *American Journal of Sociology* 95:851–87.

Marsh, Robert. 1967. *Comparative Sociology: A Codification of Cross-Sectional Analysis*. New York: Harcourt Brace Jovanovich

Marx, Karl. 1976. *Capital: A Critique of Political Economy*. New York: Penguin and New Left Review.

Massey, Douglas, and Nancy Denton. 1993. *American Apartheid: Segregation and the Making of the Underclass*. Cambridge, Mass.: Harvard University Press.

McCall, George, and Jerry L. Simmons. 1969. *Issues in Participant Observation*. Reading, Mass.: Addison-Wesley.

Mead, Margaret. 1961. *Coming of Age in Samoa: A Psychological Study of Primitive Youth for Western Civilization*. New York: Morrow.

Merton, Robert K. 1973. *The Sociology of Science: Theoretical and Empirical Investigations*. Chicago: University of Chicago Press.

Michels, Robert. 1959. *Political Parties: A Sociological Study of the Oligarchical Tendencies of Modern Democracy*. New York: Dover.

Mills, C. Wright. 1959. *The Sociological Imagination*. New York: Oxford University Press.

Moore, Barrington, Jr. 1966. *Social Origins of Dictatorship and Democracy: Lord and Peasant in the Making of the Modern World*. Boston: Beacon.

Moskos, Charles C., and Frank R. Wood. 1988. *The Military: More Than Just a Job?* Washington, D.C.: Pergamon-Brassey's International Defense Publishers.

Musheno, Michael C., Peter R. Gregware, and Kriss A. Drass. 1991. "Court Management of AIDS Disputes: A Sociolegal Analysis." *Law and Social Inquiry* 16:737–76.

Nichols, Elizabeth. 1986. "Skocpol and Revolution: Comparative Analysis versus Historical Conjuncture." *Comparative Social Research* 9:163–86.

Page, Benjamin I., and Robert L. Shapiro. 1991. *The Rational Public*. Chicago: University of Chicago Press.

Paige, Jeffrey M. 1975. *Agrarian Revolution: Social Movements and Export Agriculture in the Underdeveloped World*. New York: Free Press.

Polya, George. 1968. *Patterns of Plausible Inference*. Princeton, N.J.: Princeton University Press.

Przeworski, Adam, and Henry Teune. 1970. *The Logic of Comparative Social Inquiry*. New York: Wiley.

Ragin, Charles C. 1987. *The Comparative Method: Moving Beyond Qualitative and Quantitative Strategies*. Berkeley: University of California Press.

———. 1991. "Introduction: The Problem of Balancing Discourse on Cases and Variables in Comparative Social Research." In *Issues and Alternatives in Comparative Social Research*, edited by Charles C. Ragin, 1–8. Leiden: E. J. Brill.

Ragin, Charles C., and Howard S. Becker. 1992. *What Is a Case? Exploring the Foundations of Social Inquiry*. New York: Cambridge University Press.

Robinson, W. S. 1951. "The Logical Structure of Analytic Induction." *American Sociological Review* 16:812–18.

Rokkan, Stein. 1970. *Citizens, Elections, Parties*. Oslo: Universitetsforlaget.

———. 1975. "Dimensions of State Formation and Nation-Building: A Possible Paradigm for Research on Variations with Europe." In *The Formation of National States in Western Europe*, edited by Charles Tilly, 562–600. Princeton: Princeton University Press.

Rothschild, Joseph. 1981. *Ethnopolitics: A Conceptual Framework*. New York: Columbia University Press.

Rueschemeyer, Dietrich, Evelyne Huber Stephens, and John D. Stephens. 1992. *Capitalist Development and Democracy*. Chicago: University of Chicago Press.

Schermerhorn, R. A. 1978. *Comparative Ethnic Relations*. Chicago: University of Chicago Press.

Schwartz, Howard, and Jerry Jacobs. 1979. *Qualitative Sociology: A Method to the Madness*. New York: Free Press.

Scott, James C. 1976. *The Moral Economy of the Peasant: Rebellion and Subsistence in Southeast Asia*. New Haven: Yale University Press.

———. 1990. *Domination and the Arts of Resistance: Hidden Transcripts*. New Haven: Yale University Press.

Shaw, Clifford. 1930. *The Jackroller*. Chicago: University of Chicago Press.

Simmel, Georg. 1950. "Dyads and Triads." In *The Sociology of Georg Simmel*, trans. by Kurt Wolff, 122–69. Glencoe, Ill.: Free Press.

Skocpol, Theda. 1979. *States and Social Revolutions: A Comparative Analysis of France, Russia, and China*. New York: Cambridge University Press.

———. 1984. "Emerging Agendas and Recurrent Strategies in Historical Sociology," In *Vision and Method in Historical Sociology*, edited by Theda Skocpol, 356–91. New York: Cambridge University Press.

Smith, Dorothy E. 1987. *The Everyday World as Problematic: A Feminist Sociology*. Boston: Northeastern University.

Smith-Lahrman, Matthew. 1992. "Coffee House Cotillion: The Construction of Private Space in a Public Place." Unpublished manuscript, Department of Sociology, Northwestern University, Evanston, Ill.

Stack, Carol B. 1974. *All Our Kin: Strategies for Survival in a Black Community*. New York: Harper & Row.

Stephens, Evelyne Huber. 1989. "Capitalist Development and Democracy in South America." *Politics and Society* 17:281–352.

Stephens, John D. 1979. *The Transition from Capitalism to Socialism*. Urbana: University of Illinois Press.

Stinchcombe, Arthur L. 1968. *Constructing Social Theories*. New York: Harcourt, Brace, Jovanovich.

———. 1978. *Theoretical Methods in Social History*. New York: Academic Press.

Stinchcombe, Arthur L. et al. 1980. *Crime and Punishment—Changing Attitudes in America*. San Francisco: Jossey-Bass.

Strauss, Anselm. 1987. *Qualitative Analysis for Social Scientists*. New York: Cambridge University Press.

Suchar, Charles S., and Richard A. Markin. 1990. "Forms of Photo-Elicitation: Narrative Reflections of Micro-Social Realities." Paper presented at the annual meeting of the International Visual Sociology Association, Whittier College.

Tilly, Charles. 1984. *Big Structures, Large Processes, Huge Comparisons*. New York: Russell Sage Foundation.

Truzzi, Marcello. 1974. *Verstehen: Subjective Understanding in the Social Sciences*. Reading Mass.: Addison-Wesley.

Tukey, J. W. 1977. *Exploratory Data Analysis*. Reading, Mass.: Addison-Wesley.

Turner, Ralph. 1953. "The Quest for Universals in Sociological Research." *American Sociological Review* 18:604–11.

Vaughan, Diane. 1986. *Uncoupling: Turning Points in Intimate Relationships.* New York: Oxford University Press.

Wade, Robert. 1988. *Village Republics: Economic Conditions for Collective Action in South India.* New York: Cambridge University Press.

Wallerstein, Immanual. 1974. *The Modern World System: Capitalist Agriculture and the Origins of the European World Economy in the Sixteenth Century.* New York: Academic Press.

———. 1979. *The Capitalist World Economy.* New York: Cambridge University Press.

Walton, John. 1991. *Western Times and Water Wars: State, Culture, and Rebellion in California.* Berkeley: University of California Press.

———. 1992. "Making the Theoretical Case." In *What Is a Case? Exploring the Foundations of Social Inquiry* edited by Charles C. Ragin and Howard S. Becker, 121–38. New York: Cambridge University Press.

Walton, John, and Charles Ragin. 1990. "Global and National Sources of Political Protest: Third World Responses to the Debt Crisis." *American Sociological Review* 55:876–90.

Weber, Max. 1949. *The Methodology of the Social Sciences.* New York: Free Press.

———. 1978. *Economy and Society,* edited by Guenther Roth and Claus Wittich. Berkeley: University of California Press.

Wieviorka, Michel. 1988. *Sociétés et terrorisme.* Paris: Fayard.

———. 1992. "Case Studies: History or Sociology." In *What Is a Case? Exploring the Foundations of Social Inquiry,* edited by Charles C. Ragin and Howard S. Becker, 159–72. New York: Cambridge University Press.

Wilson, William J. 1980. *The Declining Significance of Race: Blacks and Changing American Institutions.* Chicago: University of Chicago Press.

———. 1987. *The Truly Disadvantaged: The Inner City, the Underclass, and Public Policy.* Chicago: University of Chicago Press.

Wright, Erik O. 1985. *Classes.* London: Verso.

Zablocki, Benjamin David. 1980. *The Joyful Community.* Chicago: University of Chicago Press.

Glossary / Index

A

B

C

Configurations are combinations of characteristics or aspects of cases. When social researchers view cases in terms of their different combinations of similarities and differences, they study configurations.

Constant comparative method is a general technique used by qualitative researchers to aid the formulation and clarification of concepts in the process of collecting data. It is broader and more inclusive than analytic induction, but like the latter involves an ongoing dialogue of ideas and evidence. 93

Correlation refers to the degree of covariation between two variables, usually across a large number of cases. When two variables, like years of education and subsequent income are correlated, their values correspond systematically, at least in a rough way, across cases.

Correlation coefficients such as Pearson's *r* provide exact quantitative assessments of the strength of a pattern of covariation between two variables. They vary between –1.00 and +1.00, that is, between perfect negative correlation and perfect positive correlation.

Covariance refers to the mathematical quantity that results from multiplying and then averaging deviation scores. It is similar in some ways to the calculation of the correlation coefficient, but does not vary between –1.00 and +1.00. 170

Covariation refers to a pattern of correspondence between two variables. If two variables covary, then particular values on one variable tend to be associated in a systematic way with particular values on another variable.

Culturally or historically significant phenomena are events, trends, or processes that are relevant to members of society or some other social group because of their relevance to the identity or the history of this group—who they are and how they became who they are.

D

E

Experiments occur primarily in laboratories. Usually, the impact of some test condition is compared in a controlled way across identical situations. For example, a chemist might examine the impact of different amounts of heat on some substance. 14

Explanations. *See* Causal explanations

F

Feagin, Joe R., 101

Fieldwork, 13

Fixed frames are most common in quantitative research where the goal is to test a particular hypothesis. When analytic frames are fixed, the relevant cases and aspects of cases (variables) change little, if at all, over the course of the investigation. 74

Flexible frames are most common in comparative research, especially when researchers want to assess diversity across a range of cases. Flexible frames work best when the researcher believes that there are several types of cases among those relevant to a study and wants to specify the types in the course of the research. 74–75

Fluid frames are most common in qualitative research, especially when the goal is to study cases in an in-depth way in order to clarify concepts and categories. When frames are fluid, the investigator may not finish framing by case and framing by aspect until the research is complete. 75–76

Formal properties are aspects of social units that are generic in the sense that they transcend particular settings. They include attributes like size and hierarchy and patterns of association like dyads and triads. 9

Framing by aspect involves specifying the main features or attributes of a particular category or set of cases, usually based on the general ideas associated with a social theory or perspective. 64–66, 72. *See also* Analytic frames

Framing by case involves specifying the major category or set relevant to an investigation, using theory, knowledge, and general ideas as guides. 63–64, 72. *See also* Analytic frames

Frank, Andre Gunder, 36

G

General patterns
- accumulated knowledge of, 136
- identification of, 34–35, 133–134
- research approaches to identify, 51–52

Generality is used by social researchers to refer to the breadth of the application of a concept, idea, or relationship. The greater the number and diversity of relevant cases, the greater the generality. 137

Generalizable knowledge, 34–35

Generalizations, 3

Glaser, Barney G., 91, 93, 98, 101

Global culture, 42

Goffman, Erving, 92, 100

H

Hanson, Norwood Russell, 72

Harper, Douglas, 8, 16, 44, 159

Hierarchies, 9

Historically significant phenomena. *See* Culturally or historically significant phenomena

Hochschild, Arlie, 5, 82, 90, 157–158

Hoover, Kenneth R., 3

Human Relations Area File (HRAF) is a data archive put together primarily by anthropologists that contains information, in the form of a data matrix, on many different small-scale cultures. It is an important reservoir of information on sociodiversity. 42

Hypotheses are specific propositions or "educated guesses" regarding what researchers expect to find in a body of evidence, based on their substantive and theoretical knowledge. In standard applications of the scientific method, hypotheses are tested with data specifically collected for the hypotheses.
- formulation of, 14
- testing, 15–16, 36–37
- testing social theories based on, 45–47
- theoretically based images to derive, 134–135
- *See also* Prediction

I

Idealization is a process whereby images are synthesized or built up from evidence. In social research the images that researchers construct are considered ideal in the sense that they are abstractions based on evidence from empirical cases. 70–71

Ideas most often appear in social research as explicit or implicit social theories about social phenomena. Generally, social researchers participate in long-term, ongoing debates and conversations about social life, which provide the background for the development of specific theories. 55

Images are what researchers construct when they try to make sense of or synthesize their evidence. Generally, the construction of images involves the linking of different pieces of evidence, which, in turn, exist because of the application of an analytic frame to data. Different research methods shape how researchers link bits of evidence together to form images.

Imaging is the process of constructing images from evidence and complements the process of framing by case and by aspect. 72

Independent variables are also known as causal variables. When one variable is used to explain or account for the variation in another variable, it is called causal or independent. Variation in levels of nutrition, for example, may be used as an independent variable to account for variation in average life expectancy across countries.

Induction is the process of using evidence to formulate or reformulate a general idea. The process of constructing images (via the synthesis of evidence) is mostly inductive. Generally, whenever evidence is used as a basis for generating concepts, as in qualitative research, or empirical generalizations, as in quantitative research, induction has played a part. 15

J

K

L

M

Marginalized groups are usually groups in society that are out of the mainstream or who have been pushed out of the mainstream. Generally, they "lack voice" in society because their views and experiences are only occasionally represented in the popular media, if at all. Social researchers may study marginalized groups because they have not been represented or because they have been seriously misrepresented, by the popular media or by other social researchers.

Measures are implementations of variables in a particular set of data. Generally, every variable may be measured in a variety of ways, and researchers must justify the specific measures they use for each variable. 11–12

N

Negative correlations exist when high values on one variable tend to be paired with low values on another variable and vice versa. For example, rates of literacy and infant mortality are two variables that are negatively correlated across countries. 146–148

O

P

Parsimony in quantitative social research refers to the use of as few independent variables as possible to explain as much of the variation in a dependent variable as possible. 137

Positive correlations exist when high values on one variable tend to be paired with high values on another, and low values tend to be paired with low values. For example, nutrition levels and average life expectancy are two variables that are positively correlated across countries. 146–148

Prediction is the use of accumulated social scientific knowledge about general patterns and past events to make projections or extrapolations about the future and other novel situations. Generally, social researchers can make projections about rates and probabilities, but not about specific events, like the timing of a major political change.

Presence-absence dichotomies are variables that have two values, one indicating that a condition or feature is present, the other indicating that it is absent. The comparative analysis of configurations focuses on combinations of presence-absence dichotomies.

Q

Qualitative research is a basic strategy of social research that usually involves in-depth examination of a relatively small number of cases. Cases are examined intensively with techniques designed to facilitate the clarification of theoretical concepts and empirical categories.

Quantitative research is a basic strategy of social research that usually involves analysis of patterns of covariation across a large number of cases. This approach focuses on variables and relationships among variables in an effort to identify general patterns of covariation.

R

Random samples are often used when researchers have many more potential cases or observations than they need and can afford to collect. Investigators use systematic procedures to select a representative subset of all potential cases. For example, a researcher might select every 100th case from a complete listing of all cases.

Relationships exist among variables when they covary in some systematic way. Quantitative social researchers focus on relationships among variables when they conduct research.

Reliability concerns how much randomness there is in a particular measure. When there is a lot of random error in a measurement procedure, the resulting measure is considered unreliable and successive measurements applied to the same cases will not be strongly correlated.

Representations of social life are theoretically structured and informed descriptions of empirical phenomena. In social research, representations incorporate both theoretical ideas and systematic evidence about the phenomena.

Representativeness concerns the degree to which the cases included in a study resemble either the larger set from which they are drawn or the larger set that the researcher wishes to generalize to.